Electros

Exploring, Controlling and Using Static Electricity

Second Edition

including

The Dirod Manual

A. D. Moore

Laplacian Press
Morgan Hill, CA

Copyright © 1997 by Electrostatic Applications. All rights reserved.

Published by Laplacian Press
A Division of Electrostatic Applications
16525 Jackson Oaks Drive
Morgan Hill, California 95037-6932, USA
Telephone: (408) 779-7774   Fax: (408) 779-3638
Email: electro@electrostatic.com   Website: http://www.electrostatic.com

First edition originally published by Anchor Books, Doubleday & Company, Inc.,
Garden City, New York ©1968

Publisher's Cataloging in Publication Data

Moore, A. D. (Arthur Dearth) 1895-1989
Electrostatics, Exploring, Controlling, and Using Static Electricity, 2nd edition.
First edition published: Garden City: Anchor Books, Doubleday and Company, Inc.
Includes bibliographical references and index.

1. Electrostatics
I. Title II. Author

QC571.M663   537.1
ISBN 1-885540-04-3 (printed on acid-free paper)   96-080313
                                                              CIP

Transactional Reporting Service
Authorization to photocopy items for internal or personal use, or the internal or personal
use of specific clients, is granted by Electrostatic Applications, provided that the appro-
priate fee is paid directly to Copyright Clearance Center, 222 Rosewood Drive, Danvers
MA 01923, USA. Members of the ESA may make copies for their private use without
paying fees.

Academic Permissions Service
Prior to photocopying items for educational classroom use, please contact the Copyright
Clearance Center, Customer Service, 222 Rosewood Drive, Danvers MA 01923, USA.
(508) 750-8400.

2 3 4 5 6 7 8 9 0
Printed in the United States of America

Contents

Part 1
Electrostatics

Part 2
The Dirod Manual

About the Author

A. D. MOORE was only six years old when he announced to his family that he would be "a teacher and an electrical engineer." Both of these predictions came true. What he did not foresee was that he would also have rewarding careers as an inventor, public servant, lecturer-demonstrator, and author. Professor Moore (he was called just "A.D." by almost everyone who knows him) was born on January 7, 1895, in Fairchance, Pennsylvania, and it was there that his wide range of interests and his insatiable curiosity became manifest. During his childhood this curiosity developed into an urge to experiment and invent things and to build his inventions with his own hands. As a teacher, Professor Moore often used his own hand-built devices for demonstration purposes.

Professor Moore earned his degree in electrical engineering in 1915, when he was graduated from Carnegie Institute of Technology as an honor graduate. The following year he joined the teaching staff of the University of Michigan and commenced a teaching career there that was to last for more than forty-seven years. During his tenure, Professor Moore received his M.S.E. degree from that university (in 1922), and at his retirement he became professor emeritus.

In 1940 A. D. Moore began a secondary career as a lawmaker when he was elected to serve on the city council of Ann Arbor, Michigan. This was the beginning of seventeen years of continued service on the council, and in his next to last year he was elected council president, serving as acting mayor in the mayor's absence.

There were other interests, too, to occupy this busy man's leisure time. He enjoyed writing short stories and poems to amuse his friends and family, and playing three-cushion billiards amounted to something of a passion with him for many years. He became quite expert at the latter (his high run is eleven) and several times participated in exhibition games with Willie Hoppe, the world champion at three-cushion at the time, and the great trick-shot artist, Charlie Peterson. One of these games was presented to a crowd of 2200 at the U. S. Naval Academy at Annapolis.

Retirement was anything but quiet for Professor Moore. After he began his retirement furlough in 1963, he was guest lecturer at many universities and covered over 150,000 "station wagon miles" lecturing to high school science clubs and college students all over the United States, Canada, and Europe and demonstrating his electrostatic generators and fluid mappers—his own term for the simulation of fields by fluid flow. He continued to write and present his "Science Vignettes," which numbered over 100, for the Science Club of the University of Michigan. The University's Dept. of Gerontology honored him with a plaque which stated, "...as a dedicated citizen, teacher...you continue to inspire all...as a retiree who has never retired, you exemplify the life long pursuit of laudable goals..." He maintained his office-laboratory on the Michigan campus, where he discovered and developed a new line of phenomena in the field of electrospherics and magnetospherics.

His own work in the field of man-made electrostatics led Professor Moore to the conclusion that, although a great deal of research is being conducted in the area of natural electrostatic phenomena, much more broad-front research in man-made electrostatics is vitally needed. Professor Moore worked actively toward the establishment of an electrostatics research center to fill this need. In 1970 he was instrumental in the founding of the Electrostatics Society of America (ESA) and served as the first president.

Professor Moore served as president of the national council of Tau Beta Pi from 1924–30 where he introduced the Tau Beta Pi Fellowship Program and served as fellowship director for several years. He also helped organize the National Trichinosis Conference, served as vice president at the first two national conferences, and was vice chairman of the conference's continuing committee. For some twenty years he was head mentor at the University of Michigan and ran the Mentor System; at various times he had been president of the University Club, the Quadrangle Club, and Science Research Club; he was a member of Sigma Xi, Phi Kappa Phi, Tau Beta Pi, and other societies.

He was a fellow of the American Institute of Electrical Engineers (AIEE); he was made an eminent member by Eta Kappa Nu, the electrical engineering society; and was on the Speakers Bureau of the Detroit Edison Company. He was the first recipient of the University of Michigan College of Engineering Stephen S. Attwood Distinguished Achievement Award for outstanding achievement in engineering teaching, research, publications, and service. He was a fellow of the Institute of Electrical and Electronic Engineers (IEEE) and is listed in Who's Who in American Men and Women of Science.

Professor Moore is the author of: *The Fundamentals of Electrical Design; Heat Transfer Notes for Electrical Engineering; Fluid Mapper Patterns*; *A Fluid Mapper Manual*; *Invention, Discovery and Creativity;* and *Electrostatics, Exploring, Controlling and Using Static Electricity.* In addition, he edited *Electrostatics and Its Applications.* Among the scientific journals to which he contributed are: *Electrical Engineering (AIEE); Journal of Applied Mechanics; Annals of the New York Academy of Sciences; Journal of Applied Physics; IEEE Transactions; The Journal of the American Medical Association; The Auk; The Wilson Bulletin; Scientific American; and The American Journal of Physical Medicine.*

On October 21, 1989, at the age of 94, A. D. Moore passed away at his home in Ann Arbor, Michigan.

Publisher's Foreword

A. D. Moore's *Electrostatics* has been one of the most sought-after books in the field since it first appeared in 1968. It is written at a level that a middle school student can understand, but the scientific insights and experimental techniques that he describes also make it valuable to graduate students and professors. It is especially prized for its numerous demonstrations of electrostatic phenomena, some of which have never been fully explained.

For fifteen years after writing *Electrostatics*, A. D. Moore continued to work on his Dirods and demonstrations, seeking to simplify their construction and to make them accessible to everyone. He was especially interested in Science Fairs, and took his experience as a judge in the Canadian and United States national science fairs back to his laboratory to help youngsters build and experiment with electrostatics. Responding to numerous requests, he wrote down the hints and techniques that he worked out (as The Dirod Manual), and distributed them informally to his colleagues to help them in preparing demonstrations.

Until now, this wealth of practical advice was passed along from one experimenter to another, but was never formally published. We are happy and proud to have the opportunity to present it here for all interested in experimenting with electrostatics. Both volumes, *Electrostatics* and *The Dirod Manual*, are bound together in this single book.

This publication would not have been possible without the help and support of A. D.'s colleagues, friends, and family. Special thanks go to Anne S. Benninghoff, Jeanne M. Goodman, Robert Gundlach, Lance F. Jerale, Arthur Moore, Jr., Jo Moore, Glenn Schmieg, Albert Seaver, and to the Electrostatics Society of America.

Joseph M. Crowley
Morgan Hill, California
January 11, 1997

Electrostatics Society of America

Shortly after completing this book, A. D. Moore called together several dozen individuals with an interest in electrostatics to plan how they could come together regularly to exchange ideas, questions, and inventions. These meetings were so successful that a formal organization, The Electrostatics Society of America, was founded in 1968 to serve the needs of all those interested in electrostatics, regardless of individual disciplines or education level.

An annual conference of the Society (usually in June) brings together members from all over the world to present results, discuss current work, and visit electrostatics facilities. Occasionally manuscripts are published in a proceedings volume or submitted to a special issue of an international journal. The Society also publishes a bimonthly newsletter, as well as occasional monographs covering particular topics.

Many of the demonstrations and generators described in this book are still in use by members of the Society, and the lecture-demonstrations pioneered by A. D. Moore are still given on a regular basis.

The ESA also recognizes individuals who have contributed to electrostatics in special ways. The Electrostatics Hall of Fame at the Franklin Institute in Philadelphia honors the greatest contributors to the field. Annual cash prizes are also awarded to teachers and students for outstanding work in science fairs or classrooms. Nominations are made by members of the Society.

Membership ($20.00 per year) includes the newsletter, occasional monographs, and reduced registration fees at the annual meeting. Further information or application for membership can be obtained by visiting the web site:

http://eng.bu.edu/~mnh/esa.html

or by writing to the Secretary,

Dr. Emery Miller
641 East 80th Street
Indianapolis, IN 46240 USA

Author's Preface

Did the cave man have a cat? If so, he was our first electrical scientist. As he stroked his pet in a dry cave, the fur would get "charged." It would become rumpled and ridgy; and in the dark, there would be tiny sparks to be seen. As a matter of fact, he didn't need the cat, for he surely had furs, taken from animals he had killed. These would also become charged, if stroked. But, you may ask, did the cave man take notice of this? I think so. We know that he was a most acute observer of numerous natural phenomena. He had to be, in a dangerous world, in order to "make a living" and to survive while doing it. Of course he noticed the *difference*: the difference between unstroked fur and stroked fur.

Skipping down through thousands of centuries, we come to the Greeks. In their highly developed culture, their opportunities for developing the electrical sciences were vastly better than those of our cave man. They did have that natural magnet, the lodestone, and knew that it would attract iron. They did have amber, and knew that rubbing it would make it attract particles of this and that. So they, along with cave men, were observers. They *noticed* a difference. Moreover, they had ideas, they loved to erect theories, they loved to argue. Some of their ideas about natural phenomena were sound, but others were ridiculous. That is, their make-up was such that they would rather "settle" a matter by debate than *test* it. Their lack of an urge to test, to experiment, is a striking thing. Were it not for that, they might have started the development of the electrical sciences—including electrostatics—far sooner than it did get under way.

A good scientist is a good cave man: a keen observer of the unusual, the exceptional, the different. He is a good Greek: not only an observer, but a thinker, an erector of theories, an arguer. But he goes beyond what the cave man could do, and does what more of the Greeks should have done: he *tests* his ideas. He experiments. He organizes nature to make nature reveal her truths. William Gilbert was a good scientist in this sense. Physician to Queen Elizabeth, Gilbert, around 1600, became the first to do extensive research in magnetism and electrostatics. He wrote the famous *De Magnete*. Thus did the electrical sciences get their small but real beginnings.

After Otto von Guericke gave us the first electrostatic generator, a friction machine, in about 1660, various forms of friction-type generators slowly appeared. With these, numerous early scientists could perform many experiments. Some of the facts about the nature of electricity began to be established. Benjamin Franklin had only the friction machine for his experiments, and so did the greatest experimenter of all, Michael Faraday.

The year 1800 witnessed a revolutionary advance, when Volta gave the world its first source of *steady current* in ample amounts by inventing the electrochemical cell, or "wet cell." Electrochemistry and electromagnetism could have their beginnings. The electrical sciences were growing as an ever-enlarging family. Faraday's discovery in 1831 of electromagnetic induction, plus other advances he and others made, brought about electric generators and motors and electric power; the telegraph; the Atlantic cable; the telephone; electric lights; eventually, the numerous household appliances we use so much; and electronics, with its multiplicity of uses.

In the past century, interest in electrostatics did not lag, even though many of the new discoveries and applications in the other areas of the electrical sciences were so exciting, and were becoming so rapidly useful to mankind. Electrostatics even took on new life when better electrostatic generators, of the influence (or induction) type, became available. And with the advent of the famous Wimshurst generators in 1878, some of which were giants, a new surge of hope for *applying* electrostatics developed. Inventions were made and patents were taken out in the late part of the last century. They were, however, ahead of their time. Apparently, the only useful thing the Wimshurst ever did in electrostatics was to operate some early X-ray tubes.

Thereafter, interest in electrostatics *did* lag, which is one reason for this book. By 1900, it seemed that electrostatics was not going to live up to hopes. In contrast, other new and wondrously interesting and useful applications of electricity were keeping scientists and engineers busy and happy. And not only that: electrostatics acquired a bad reputation! We began to realize that its sparks could set off explosions in flour mills (dusts can explode) and gasoline tanks; a spark during surgery can, and has, caused the death of more than one patient on the operating table, when ether in the lungs has exploded. Electrostatics makes dirt stick to walls. Paper misbehaves in printing establishments and yarns give trouble in knitting mills when "static" is generated. The unexpected spark from friction makes us jump, and we always jump too late (which is bad for our self-esteem!). "Static" from any source is resented when it interferes with audio or visual reception of radio or TV sets. And nature's electrostatics, in the form of lightning, can sometimes terrorize us. Is it any wonder that electrostatics has a bad reputation? Nearly everyone now thinks of electrostatics as a naughty little fellow, hard to control, full of tricks and practical jokes, and unwilling to take its place among the *useful* electrical sciences.

Very few, even among scientists and engineers, know that in this century, electrostatics has been put to use in a number of spectacularly successful ways. You will read about these later on. Very few realize that as we learn more about controlling the whole group of electrostatic phenomena, more and more applications will come along.

And this is where you come in. As a young scientist, you can get an early start at electrostatic experimentation, have a great deal of fun, and learn a lot as you go along. Your experimentation experience alone will be invaluable to you. In college, you can round out your grasp of theory. Then, in practice, you may be one of those to make some of the advances that certainly lie ahead.

The young scientist today needs, more than ever before, to become an experimenter, for there has been a profound change in the American scene. I'll bring it out by going back to my own boyhood in southwestern Pennsylvania. I was raised on the farm, where there are many problems to be solved and many handy things to be learned. And when painters or roofers or carpenters or plumbers or threshers would come, endless questions could be asked—and I surely asked them. (Today, only 6 percent of the American population come from the farm.) But when loose from farm chores, I spent lots of time in the plumbing shop, the hardware store, the blacksmith shop, the foundry, the machine shop, the brickyard, the lumber mill, and the glass plant. There were the coal mines to visit, and the coke ovens. There was a power plant, where the engineer was my friend. There was the streetcar line, where the motorman was my friend. There was the telephone man, who came to put new dry cells in our telephone, and he would give me the old ones to use in my experiments. It was a very rich environment for a kid who wanted to be an electrical engineer. Along with all this, I loved to make things like bows and arrows, windmills, water wheels, miniature ovens, squirt guns, kites, and other important things. I had my failures and my successes. I learned a great many things about materials and how to use them.

Coming back to you: today, safety rules prevail, and you, to my great regret, cannot wander at will in a glass plant or other factory. "KEEP OUT" is a familiar sign. This denial of free access to American industry is a great loss to a full childhood, and believe me, a great loss to science and engineering. What can you do to help make up for it? You can steam ahead on your own, as an experimenter, learning about materials and processes and functions, having your own failures and successes, acquiring common sense and judgment as to what will work and what won't. There is just no substitute for acquiring this kind of know-how.

If this book does nothing more than coax you into experimentation with electrostatics—or anything else!—it will have served a good purpose.

I love to invent; develop the idea; then design the device; then build it, preferably by myself; then experiment with it, test it, perfect it, and learn from it. This is a "lab-written" book, conceived and largely written right in the midst of such activity. I have no real office, and don't want one. My desk is in my laboratory, surrounded by the various equipment I have developed. Thus, when an idea strikes, I can turn at once from desk work to experimentation or design, or go down to our shop to make something. This has happened time and time again while pushing this book along.

It is unusual for a retired professor to get to retain a fine laboratory, with full facilities, in which to keep busy and from which to go out and give lecture-demonstrations; and I take pleasure in expressing my gratitude to Professor W. G. Dow, former Chairman of Electrical Engineering, University of Michigan, and to the present Chairman, Dr. Hansford W. Farris, for making it possible.

I am deeply indebted to my colleague, Professor Edward A. Martin, for the continued interest he has taken in my developments; for his ever-willing response to my frequent need to consult him; and for his careful reviewing and checking of the whole manuscript.

My thanks go to Professor Dale Ray for checking the data and computations in the early part of the book. I appreciate the way that busy men in industry took time to correspond and furnish needed information—in particular, Myron Robinson of Research-Cottrell, Inc.; G. Hall Carpenter, President, Carpco Research and Engineering, Inc.; and Dr. Emery P. Miller, Vice President, Ransburg Electro-Coating Corp. To The Detroit Edison Company, credit for the pictures of Dirod I and Dirod II; to my close friend Eck Stanger of the Ann Arbor News, for his excellent photographs of my other apparatus. To Dr. James A. Bennett of Queen's University, Kingston, Ontario, my appreciation for his interest and helpfulness during the development of much of my apparatus when, as a doctoral candidate here, he shared space with me in my laboratory. Finally, I am most grateful to editor Bruce F. Kingsbury, whose guidance and encouragement have meant more to me than he may ever realize.

Part 1

Electrostatics

1. High Voltage

Recently, I gave one of my electrostatics lecture-demonstrations for a high school science club. My six or eight generators had all performed properly, and there had been plenty of sparks. The smoke precipitator had precipitated, the electric blower had blown, the Franklin motor had whizzed, and all the other items I like to dream up had done their stuff. At the end, as I always do, I invited the young scientists to take over. They jammed around the equipment to run it, repeat experiments, and have fun.

One young fellow was especially excited. I would guess that his room is a scientific clutter, of the kind that drives a mother out of her mind. Also, he certainly must know how to work up frictional charges, for he said, "I already have my mother on the run." Then, pointing to one of my generators, he added, "If I built one of these, I could keep her out of my room *entirely.*" This is just one of the advantages of having fun with electrostatics. Another is that, if you learn to control mothers electrostatically, you may "go in" for electrostatics, and spend your life at it. I hope you do!

These bright youngsters ask me many questions. One goes something like this: "Household appliances carry only one hundred fifteen volts, yet some people die from making contact with the wires of defective appliances; or again, if you get too close to a fifty-thousand volt power line, it kills you; then why is it that your generators, or our high school's Van de Graaff, at even higher voltages, aren't dangerous?" These are mighty important questions. They deserve serious attention and some straight answers.

Electrical Hazards

Electricity is so universally used that everyone should know about its hazards. Let me tell you about a sad electrical death. I was on our city council at the time, and was making many visits to the police department. I shall never forget being there on a beautiful Sunday morning when the emergency call came in. I drove to the address where the police car went to answer the call. Inside that little home was tragedy. While the young husband had been reading downstairs, his wife was slowly dying upstairs, in the bathtub. A portable radio on a shelf over the tub, plugged into the 115 volt supply, had fallen into the water. Very probably some of the current passing through her had only paralyzed her muscles at first, so that she could not move or speak. Sooner or later breathing action, or heart action, or both, stopped. (For years, I have told my electrical engineering students that there should never be an outlet in the bathroom. I still think so.)

If the body gets a good-enough *contact* with a supply of 115 volts, and if enough of the alternating current flows through the heart, it can mean death. The heart itself is timed and triggered electrically: special cells in its "pacemaker" initiate each beat. A foreign current (and the usual alternating current of 60 cycles per second is especially deadly) can so upset the self-regulating action of the heart that either it stops, or else goes into a useless, squirming contraction called *fibrillation* and fails to pump blood.

Now, did you notice that above, I said *contact*? A 115 volt line will never reach out for you with a spark or an arc. You leave it alone and it will leave you alone. And the moral is, LEAVE IT ALONE. Do not make contact with live wires carrying 115 volts. Since I said you deserve straight answers, here comes one. The truth is that 115 volts is *relatively* safe. Any number of people in good health have accidentally (or foolishly) taken this risk and lived to tell about it. But it *can* be a killer, and especially for the very young. Babies die every year from sticking fingers into live sockets or biting through insulation on extension cords. Let it alone.

The higher the voltage, the more eager a live wire is to "reach for you" without waiting to be touched. A 50,000 volt power line would spark over to you *before* you touch it. In any case, contact or near-enough approach would guarantee anything from grave shock and burns, to death. The power generators back of that line (and back of your 115 volts in the house) keep right on turning; they continue to keep the line charged, and the deadly current continues to flow.

By the way: if you ever see a child flying a kite near a high-voltage line, step right in and take charge. A wet string might lead to a fatality.

Discharges from Small Electrostatic Generators

Tabletop models of Wimshurst and Van de Graaff generators have been on sale for years. Hundreds, perhaps thousands of high schools and colleges have them. Many young scientists have bought their own and experimented with them. Inevitably, thousands of youngsters have taken discharges from these high-voltage generators, intentionally or accidentally, and thousands more will do so. Is this deadly dangerous business? No. Risky? No. Why not? Before we get into that, let us realize a hard fact: a manufacturer who sold large numbers of hazardous devices would soon be out of business. Damage suits can be very costly.

When a Van de Graaff or a Wimshurst or one of my generators (shown in this book) tosses a spark at you, it is a spark of extremely short duration; the spark oscillates at high frequency; and the very act of sparking largely *discharges* the generator. Thus, the generator does not and cannot continuously pour energy into you. Furthermore, all of these generators are built with such a low ability to store up charge prior to the spark (we call this ability *capacitance*—to be fully explained later) that the energy you get is completely harmless. Even if it makes the muscles jerk, and a shock is felt, the energy is far, far below the danger level.

To go into it further, let's talk for a moment about big cannon balls and extremely tiny bullets. A 6-inch cannon ball is a massive thing. Fired at you from a cannon, it will wipe you out. In fact, it doesn't need to go nearly that fast. If it rolls off a house roof onto your head, you are ruined. In contrast, think of a tiny lead bullet no bigger than a grain of dust. We could fire it at your bare skin ten times as fast as a rifle bullet goes, and you might not even know it. So: the *energy* delivered to you is concerned with both *velocity* and *mass*. It is the energy that does the damage.

Likewise with these generators. How much *energy* can they store up, to be delivered by way of a spark? Very little indeed. Now, we can increase the energy stored, simply by connecting a *capacitor* to the generator, and making the generator charge *it* along with *itself*. Capacitors used to be called *condensers*, for, like the old-time Leyden jars, they "condensed" electrical energy. Later on, I will tell you how you can walk into your own kitchen, take two plastic vessels and some water, and in one minute have a dandy capacitor. If at the same voltage the capacitor can store up five times the energy that the generator can store up, the discharge would be that much more vigorous.

This brings us back to hazards again. And some *very* straight talk. Today, a wealth of materials is available, not only in stores and in scientific supply houses, but even in the home—forms of insulation such as plastic sheets, and forms of conductors such as aluminum foil. The old experimenters would have loved and cherished these materials. These and many others—epoxy glue, for instance—make it a wonderful world for today's young experimenter. But tread carefully. If you don't know what you are about, you might put together a capacitor large enough to be deadly. Therefore, later in the book, I will tell you what the known limits of safety are. As a responsible young scientist, I am sure you will stay far inside those limits. I will even tell you how to figure the value of simple capacitors, so that you will know what you are doing.

One maker of a Van de Graaff warns that you are not to connect his machine to any extra capacitor. I will describe capacitors safe to use with my generators.

2. Frictional Electricity

We live in an electrostatic world, and friction helps to make it so. There was friction when you picked up this book, and you generated some frictional electricity. When you turned to this page, you generated some more. When you shifted in your seat just then, your clothing made frictional contact with the seat, and it happened again. I invite you right now to lay the book down, take an ordinary sheet of paper, place it against some vertical surface (the plaster wall, or a wooden door), rub it vigorously with your hand—and then watch it stick right there for a while. Charges were developed, and forces of attraction are holding the paper against the wall.

Years and years ago, as a boy on the farm, I had fun making a generator out of the family cat. On a cold winter day when the air was dry, the cat and I would settle down on the floor in front of the fire in the grate. Soon she would be nice and comfy, and her fur would be warm and dry. As I stroked, she purred. But pretty soon, when the little sparks began to crackle and pop among the hairs, the purring stopped. A few more strokes; then I would bring a knuckle close to her tender nose. A little spark would jump the gap, which I thoroughly enjoyed, but she didn't. Who knows? Maybe this is why I went in for electrical engineering!

When lecturing in Ottawa recently, a professor I met, now nearing retirement, told me this very funny story on himself. Born and raised in Europe, he took his degree there, and then had extensive electrical laboratory experience. Then he crossed the Atlantic and joined a Canadian university. And in Canada, he began to notice something new in his personal life. It kept on happening, and it finally worried him into going to a doctor about it. It turned out to be "incurable." His "disease" was electrostatics! In the part of Europe where he had lived, the weather was never dry enough to let people work up charges by walking on rugs and so on. In Canada, as in much of the U. S., it is often dry enough to let frictional electricity make "personal sparks" when metal objects are about to be touched—which takes us back to the cave man in the Preface. Was his cave dry enough to let him observe frictional electricity? If not, he certainly must have observed its effect outdoors on a cold, dry day when he rubbed his hands on a piece of fur clothing.

The World's First Generator

The Greeks, at least as far back as Thales, about 600 B.C., knew about amber and its exceptional ability to become electrified when rubbed. Then, nothing much new was learned about electricity for twenty-two hundred years, until Queen Elizabeth's physician, William Gilbert, did a lot of research in electricity and magnetism. From it, he wrote his renowned book *De Magnete* in 1600. However, even Gilbert had no machine for making electricity.

The world's first electrostatic generator came into existence around 1660, at the hands of a highly ingenious experimenter, Otto von Guericke. He was already famous for inventing the vacuum pump, about 1645, and for demonstrating the effect of air pressure on two copper hemispheres closely fitted together. When the hemispheres were evacuated by his pump, two teams of eight horses each could not pull them apart. By the way, von Guericke was no secluded scientist wishing to be left alone with his investigations. He took part in the Thirty Years' War, and served as burgomaster of Magdeburg for thirty-five years.

He made a big round sulfur ball, mounted on a shaft. When it turned, and he rubbed his rough hand against it, the first generator was born. Charges appeared, and he made some of the earliest discoveries about electrostatics. Incidentally, back in those days the facilities for making things were meager indeed. Suppose you had been von Guericke's assistant in 1660, and he said, "Young fellow, I have an idea. It calls for a nice, big, round, and very smooth sulfur ball—can you dream up an easy way to make it for me?" Could you? Von Guericke did. He had a spherical glass vessel blown. He melted the sulfur, poured it in, and let it harden. Then he cracked off the glass. Ingenious? Very. Well, that's the kind of ingenuity *every* experimenter needs to have.

Friction-type Generators

If you read about the early experimenters, you find them busy rubbing something with something else. Cat fur, flannel, silk, wool, and so on, were used to rub sealing wax, glass, metal, and other things. And they found some combinations much better than others. They also found that when two rubbed-together materials were separated, they not only were charged—they were, as we say today, *oppositely* charged. It was Charles Francois Du Fay, superintendent of gardens for the king of France, who discovered in 1733 that there were two kinds of electricity. Von Guericke was the first to get away from rubbing by hand, and the first to generate electricity by machine. In the next two centuries, various kinds of friction-type generators were developed. Some of them certainly worked very well in dry weather; but apparently they all got balky, or quit, when it was warm and humid. I know little about the actual construction of these generators. They have almost all disappeared, and you would have to hunt far and wide to find one tucked away in a back corner of a museum. No modern book I know of describes them. The *Encyclopedia*

Britannica briefly describes a couple. One would have to prowl through old books and papers to unearth their history, and I'm afraid that much of it is gone beyond recall.

The great Michael Faraday, early in the last century, had only friction machines to work with. Before that, Benjamin Franklin not only used friction generators, he even treated people's ailments by using discharges on them; and he was honest enough to say that he never knew whether those who improved got better from the treatment or from the exercise of walking to and from his home.

The Van de Graaff Generator

The Van de Graaff is a modern friction-type generator you can buy, and it nicely illustrates some principles that must be built into any such generator. First, let us consider cat skin (the hair on the skin, really) and glass and flannel. Rub the first two together and, as the old-timers knew, the cat skin would become positively charged, and the glass, negatively charged. Now rub the second two together. This time, the glass comes out positive, and the flannel negative. Many materials can be arranged in combinations like this.

The Van de Graaff generator has a base with a pulley in it, a round vertical insulating tube stuck into the base, and on top of that an aluminum ball with a pulley inside. A thin flat rubber belt rides the pulleys, with a little motor in the base driving the belt. Now, in the case above, the glass was *electronegative* to the cat skin, but *electropositive* to the flannel. Using such a combination, the one pulley is covered with wool (or something like it), and the other with some kind of plastic. The belt would be electronegative to the one and electropositive to the other. Thus the belt carries one kind of charge up and the other kind down.

All this does no good without some kind of collecting device. Therefore, a piece of thin metal with a somewhat ragged edge is placed in the ball with the edge close to the upcoming belt. It picks off the charge, feeds it to the ball, and the charge then appears on the outside of the ball. Another such collecting device is in the base. Thus the base and the ball—the *terminals*—store up opposite charges. Connect a conductor to the base, bring it near enough to the ball, and a brilliant spark will jump the gap.

Every high school should have a Van de Graaff. Smaller ones accumulate up to 200,000 volts; larger ones, much higher. For varied electrostatic experimentation, however, you need a generator with both terminals on the same level and conveniently accessible for connecting to devices. You also need speed control. Apparently, most Van de Graaffs used in high schools operate at only one speed.

The "professional" Van de Graaff is, of course, a giant, with charges sprayed onto the belt from a separate power source, and not friction-generated. It makes millions of volts for particle acceleration research.

Frictional Electricity: Troubles and Hazards

Once, driving my car in hot dry weather in the Southwest, my radio had so much static interference that it was useless. When the car stopped, the radio worked all right. The rubber tires running on the smooth dry highway were building up charges that had to discharge somehow, and the discharges caused the interference. Under such conditions, it is common to get a shock when stepping out of the car.

Long before that, the chief engineer of a company making railway signaling equipment told me what happened when they were installing a complete signal system on a western railway. One day he had a frantic telephone call: "Send somebody quick! Every light on the system is red, and the whole system is shut down!" He asked if they were having a dust storm and was told they were. Windblown dust is often highly charged, and dust particles in this case had put so much extra charge on the signal wires that safety devices automatically had worked to turn all signal lights red. A ground wire, put up to prevent just this kind of thing, had not yet been grounded. When the wire was grounded, the trouble disappeared.

An early student and good friend of mine, Dr. Ross Gunn, had a long career in the investigation of atmospheric electricity, storms, and the like, and published many papers. He has had some exciting experiences. When a military plane flies into an electrified cloud, it can pick up charges which, draining off by discharge, can ruin radio communication. During the Second World War, the government had Dr. Gunn do a lot of direct research, flying in an instrumented plane. He made many dangerous flights through storm clouds, and the plane itself was struck by lightning on three occasions. He found that dry snow, or ice crystals, sliding over a plane's metal surface develops frictional charges, with the plane always becoming negative and the particles leaving it, positive. Once, the plane developed a charge of 450,000 volts!

I have already mentioned explosions set off by electrostatic discharge, including ether explosions in the operating room. Here is a situation where one requirement interferes with another. To avoid the spread of infection, hospitals like to put a smooth floor in a surgery room, and keep it waxed and polished. Such surfaces are our worst enemies when trying to make a place "staticproof." They generate charges when walked on. It may be a long time before we are smart enough to have everything else the way we want it, and still remove all electrostatic hazards from the operating room.

What Goes On Here?

Before we close this chapter, we certainly ought to ask: How does friction make charges appear? And is it really friction, or could it be something else? We also need to get some names clarified. The early experimenters certainly called it frictional electricity. Then someone proposed that the name be changed to *triboelectricity*. When you look up *tribo*, you find it is Greek, meaning "to rub": hence, *rubbing-electricity*! For my part, I can't see that the newer name is of much help. Perhaps it is justified by the fact that the word *friction* is so frequently used to denote *mechanical friction*, in which there is a force resistance and heat is developed. This friction force, and the heat, apparently have nothing to do with making the charges appear. So next we turn to another name, *contact electricity.*

It is a fact that when almost any two solids are made to touch, a voltage difference, or *contact potential*, occurs. In most cases, it is very small; but with tin and iron, for example, it is nearly a third of a volt. The tin is electropositive in this case. Moreover, if a piece of plastic, for example, is merely pressed against a metal plate—not rubbed—and taken away, it will have a charge at any area where actual contact was made. Nowadays, it is believed that what rubbing does, and *all* that it does, is to *increase the number of little areas that actually make contact.*

Now, do you suspect what we are getting around to? All three names mean the same thing; for *contact* includes either pressing or rubbing.

Call it what you wish, for all three names may survive for a long time. But we still need to answer that question, What really happens? Well, an uncharged piece of plastic consists of *neutral molecules*, each one made of neutral atoms: the positive charges in the nucleus of the atom (called *protons*) are equal in number to the negative charges (called *electrons*) sailing around them. The metal plate, uncharged, or neutral, is made of its own kind of atoms. Now we touch the plastic and the metal together—and one of them becomes an electron thief! Somehow, it steals some electrons from the other one's surface units. It turns out in this case that the metal is the thief. When we separate the two, some of the plastic's surface molecules have lost one or more electrons each. And the plastic, being a *nonconductor*, tends to retain that state at any little area of contact. Do some rubbing now, make a great many more little contact areas, and that many more little areas become charged. Probably it is impossible to charge the whole area, or even come close to it. But it *seems* to us as if that were the case.

Otto von Guericke knew nothing about this phenomenon, no matter what name you give it. And now comes the surprise: three centuries of scientific advances have been made, and we still do not know all about it. There is just no theory that will take all of the truly remarkable knowledge we do have of the properties and behaviors of atoms and molecules—that will take this knowledge and wrap it up into an explanation that completely accounts for contact electricity.

All of which leads to the next chapter, where we must get down to business, bring out some basic truths about the atom, and get closer to an understanding of what we call charges.

3. Let's Talk About Charges

In older times, an electrical charge was a completely mysterious thing. Early experimenters, such as Otto von Guericke, Benjamin Franklin, Michael Faraday, and many others, knew that they had something, but they didn't know what it was. Calling it "electricity" gave it a name but did not explain it. They knew that somehow, friction between two different materials could produce electrical effects. They knew how to make friction-type generators to get more effect than by just rubbing things together by hand. They made sparks, and wondered about them. They knew how to "charge up" an insulated object—even themselves. They knew that objects charged alike would repel each other; and that those with unlike, or opposite charges, would attract each other. Thus they made some discoveries, but there was much left to argue about. What, really, is this thing we call *charge*?

Above, I mentioned charged *objects*. Quite possibly, you have always thought that only solid things get charged—such as yourself, or a metal plate, or a plastic raincoat. However, liquids can be charged, too. Even more: when you consider that lightning is a huge electrostatic discharge, and that an electrified rain cloud isn't very solid, and certainly isn't a continuous volume of liquid, you wonder. And when I tell you that on walking into a room, the roomful of air surrounds you with billions of charged particles, you may wonder some more. Not that there is magic in the room; step outdoors, and you are still so surrounded. You *live* in an electrostatic environment. Furthermore, many of the phenomena of electrostatics are intimately concerned with charges in the air, as we shall see later in this book.

What This Chapter Will Do for You

If you have already had physics or chemistry in high school, much of this chapter will be a review of what you already know. If not, fine! There is always a time to start, and I would be honored if this is where you make your beginning at learning something about atoms and their protons and neutrons and electrons. This beginning must be made sometime, and the sooner the better if you are going to end up in biology, or medicine, or electrical engineering or electronics, or chemistry or biochemistry or pharmacology, or physics or biophysics, or a lot of subsections of these fields. We can add astronomy to the list.

But suppose you won't go in for any of those. What about a business career, or that of a housewife? Some of your friends will be scientists or engineers, and wouldn't you like to know something of what they do and think about? Your newspaper and magazines will carry items about advances in medicine, science, space, and other things. In this life we lead, these matters crowd in upon us, and serve us, from all sides. We cannot escape them, and we shouldn't want to. Knowing something about the world of science in which we live is just as essential to our culture as knowing about politics, economics, sociology, good English, spelling, and hamburgers. And by learning about charges, we are opening the doors to a great many other interesting things!

Here's a suggestion: read this chapter straight through. Then come back to it again for review sometime, and dwell on it.

Atoms, Protons, Electrons, and Ions

Now let's plunge into the world of the atom to find out something about its makeup and character. The hydrogen atom is the simplest and lightest of them all. Nearly all of its mass is in its *nucleus*, which consists of one *proton*; the proton is one positive, or plus, or (+) charge. Sailing around it at a tremendous rate is one *electron*, far lighter or less massive than the proton; the electron is one negative, or minus, or (–) charge. These opposite charges are exactly equal, so that the atom as a whole is *neutral.*

The next possible atom up the line of *elements* is easy to dream up: it would have a nucleus of two protons and have two electrons in orbit around it. Yes, but would it be a realistic atom? Would it "work"? You at once point out that the two protons, being *like* charges, would repel each other and would refuse to "stay put"—they need a "glue" to hold them together. This atom we are building is going to be a *helium* atom; when its mass is determined, it is found to be about *four* times as much as the hydrogen atom's mass, instead of twice as much. This is because it also has two *neutrons* in that nucleus: *neutral* particles that somehow hold the two protons together. Let's go on up the line to *oxygen*, with eight protons in the atom's nucleus, eight of the neutral neutrons in the nucleus, and eight electrons sailing around in their orbits.

Two more things about oxygen: the air you breathe is 21 percent oxygen; and you are breathing *molecules*, not atoms, of oxygen. In the gaseous form, in air, the molecule consists of two atoms of oxygen and it has the symbol O_2. If, somehow, we could knock an electron out of a molecule, we would form an *electron-positive ion pair.* The free electron would start out on its own, as a free negative charge; the molecule now has one extra positive charge, thus becoming what we call a positive *ion.*

The electron, after a vast number of near-collisions in an extremely short time, would soon find a neutral oxygen molecule to *join*. This is called *elec-*

tron attachment. This molecule now has an extra negative charge and has become a *negative ion*. We now have two ions, one positive, one negative.

Now, what would knock that electron out in the first place? Two things are doing it all the time. One is *cosmic rays*: packets of enormous energy coming in from unknown sources in outer space. These rays create little showers of various particles, and one of the effects is to knock electrons out of oxygen molecules. The other cause is *radioactivity* caused by the radioactive content of the earth. The total effect is that in the lower atmosphere where most of us live, about 10 to 20 new pairs of ions are formed this way every second in each cubic centimeter of air.

If this kept on happening, and nothing else happened, all of the oxygen would soon be *ionized*. But something else does happen. The ions are forever dashing about and colliding; and when a positive ion meets a negative ion, they neutralize each other: the positive ion that needs an electron takes it from the negative ion that has the extra electron. This is called *recombination*.

The final outcome is that in reasonably clean room air or outside air, there are from 100 to 500 ions in each cubic centimeter. Verily, you do live in an electrostatic world! And one more remark: since these ions are extremely few compared to the vast number of neutral molecules they are among, two opposite ions have to hunt around quite a bit before they find each other and recombine.

Some Practice with Big Numbers

Soon we will be taking a look at a little aluminum cube and getting some facts out of it about charges and electrostatic effects. And we will come up against some numbers so tremendously huge that we had better review our knowledge of big numbers. The way to do this so that our minds can follow is to start with small things that we can see.

I have collected a great many odd things in my laboratory, for my experimentation, and one is a little bottle of plastic beads. These, roughly spherical, are each just about a millimeter in diameter. Ten of them lined up reach for a centimeter. With 100 centimeters in a meter, that makes 1000 beads per meter. Now, the short way to write 1000 is 10^3, meaning, $10 \times 10 \times 10$. The short way to write 1,000,000, or one million, is 10^6, or six tens multiplied together.

How many beads in a roomful? Take a room 4 meters square and 3 meters high (about 13 by 13 by 10 feet) and fill it with beads. Assume that the beads arrange themselves on the corners of little cubes to make our figuring easier (actually they would settle among each other somewhat). The room would hold $4000 \times 4000 \times 3000$ beads, or 48×10^9; or, written out, 48,000,000,000 beads—in words, forty-eight thousand million, or forty-eight billion beads.

You and I can easily imagine that room, and easily imagine it full of beads. But can we really *comprehend* that *number* of beads?

Well, get ready to stretch the imagination until it cracks. Think of a one-story house made up of sixteen million of these rooms, *all* filled with beads. That would be nearly 7.7×10^{23} beads, or about 770,000,000,000,000,000,000, 000. This isn't play. This is real. We will soon need that large a number.

A Little Aluminum Cube

When I lecture on electrostatics, I show my little aluminum cube and talk about it, as I will do right now. It is a "centimeter cube"; it has a volume of one cubic centimeter. Now, aluminum is that element whose atom has 14 neutrons and 13 protons in the nucleus, and 13 electrons around the nucleus. Unlike the wandering molecules of the air, these atoms have to stay relatively fixed.

Yes, they vibrate and do other odd things, but they are not allowed to leave home.

How many atoms are there in the cube? When you learn to use Avogadro's number, you will easily figure it out. Using it, we find that the cube is made of 0.602×10^{23} atoms. If you multiply that by 13, for the number of electrons in an aluminum atom, you get 7.8×10^{23} electrons—that is, 780,000,000,000,000,000,000,000 electrons. Now then, how many of those plastic beads did we have in sixteen million roomsful of beads? We had 770,000,000,000,000,000,000,000, or about the same number. Fantastic, isn't it?

If we could take these electrons out and lay them on a line one centimeter apart, how far would they reach? Around the earth? That isn't even a start. Let's lay them out to the sun—93,000,000 miles and back. Even that makes hardly an impression. To get rid of them all, it would take 26,000,000,000 *round trips to the sun*. Again, fantastic? (Just for fun, you might find how many round trips it would take going to the nearest star, four light-years away.)

Our limited minds can never conceive of how small and how numerous atoms and electrons are; but calculations like these do help.

An Enormous Force

Unlike, or opposite, charges attract each other. Imagine, with me, that we could take all of the electrons out of that little cube and move them to a point 1 meter away. The cube now has only positive charges—protons. The electrons are all negative charges. What would be the force of attraction between the two groups of charges? Look out—here comes another big number. It turns out to be 32×10^{18} pounds, or thirty-two million million million pounds! To get at least a vague idea of that force, it is equal to the weight of a steel cube 76 miles high! If you started driving from one corner of it at 76 miles an hour, it would take an hour to reach the next corner.

When you think of how small the aluminum cube is and how enormous that force of attraction is, you can easily get the idea that very large electrostatic forces can somehow be made to appear and be put to practical use. That is exactly the *wrong* lesson to learn. What we have really learned from this example is that those electrons do not want to leave home, being powerfully attracted by the protons around which they are whizzing. *Useful* electrostatic forces are actually *little* forces, not big ones. An electrified comb can pick up pieces of paper, but cannot pick up the family cat. As we shall see later, electrostatic forces can do a superb job at moving smoke and dust particles and paint spray particles and some larger particles. And of course, they can be made to wiggle a beam of electrons back and forth to make your TV set come alive. Little forces, yes. Big forces, no.

What Is Charge?

We have seen that the smallest electrical charge in our universe is the negative electron, or its opposite mate, the proton. When these are present in equal numbers in a thing, whether the thing is an atom, a molecule, the volume of a solid, or even a volume of gas, that thing is neutral.

When a *conducting solid* becomes charged, it has somehow gained or lost electrons. Carefully note this: its *interior* remains neutral. The extra charge appears on the *surface* and spreads all over it. How it spreads, that is, how it becomes *distributed*, will depend on the shape of the solid, and the presence of nearby objects and charges, and so on.

When an *insulating solid* becomes charged, the charge tends to remain anchored to the area where it was developed. Good insulators can hold the charge for quite a while. Poor insulators let the charge leak to other places; also, surface contamination on even a good insulator may make the surface somewhat conductive, and let leakage occur.

A *volume of space* can become charged. Suppose I walk into my laboratory on a day when the space in it happens to be filled with neutral air. Of course, nearly all of the molecules are always neutral. However, cosmic rays and radioactivity are busy, and there are some positive and negative ions; but they are equal in number this time, and the space has no charge.

But now I start one of my generators, connect a sharp point to its negative side, and make some negative *corona* (we will have much to do with corona later on). From the negative corona comes a flood of negative ions. These ions, in the space near the corona and spreading out from it, give that *space* a charge, and this is called a *space charge*. Or again, consider a thundercloud, most of the bottom of which is usually negatively charged. Here is nature's way of exhibiting a space charge on a grand scale.

4. The Dirods: Induction-type Generators

Friction-type generators were finally replaced by generators of an entirely different kind, but it took a long time to do it. The *Encyclopedia Britannica* gives a brief history of electrostatic machines, and it credits Abraham Bennet (inventor of the gold-leaf electroscope) with devising the first generator of the new kind: Bennet's doubler. This was in 1787. Instead of making charges appear by friction, Bennet caused *induced* charges to make his machine generate. We shall very soon see what induced charges are, and how they can be used. Bennet's machine, and those that followed, were called *influence* machines. The word "influence" is no longer used in discussing electrostatics in general, and I prefer to call them *induction* machines. Now, induction of charge had been discovered by John Canton as early as 1753. Thus, a third of a century went by before Bennet put it to use.

Another form of doubler was brought out by William Nicholson in 1788; then Belli's doubler appeared in 1831. Around 1860, Lord Kelvin was using similar (very small) devices in his electrometers. He called them "replenishers." The first really successful high-voltage induction generator came with Varley's machine in 1860. One wonders why it took seventy-three years, from Bennet to Varley, for the new generator to grow up. Varley's generator was soon followed by August Toepler's; then, from 1864 to 1880, Wilhelm Holtz produced a variety of machines. Finally, James Wimshurst came up with his extremely ingenious form of induction machine, in 1878 and after—and it took the field. In a Chicago museum, I have seen a giant Wimshurst. Its two round glass disks, turning in opposite directions, are 3/8 of an inch thick and 7 feet in diameter!

Next, we look at induced charges in a generator.

Induced Charges

The first rotating electrostatic generator I designed (see Plate 1), is now called Dirod I. If you could walk into my laboratory (and how I wish you could!) we would go directly to this induction-type machine, start it up, and at once have it throwing a steady stream of little sparks to your knuckles. As you see, it has a lot of rods, stuck through holes in a disk. I coined the name Dirod from *disk* and *rod*. To see why it works, look at Figure 1. Here we see only the electrical parts of a simple six-rod machine.

Plate 1. Dirod I, an induction-type electrostatic generator with 36 rods.

The *disk*, D, is made of Plexiglas, a high-grade insulator (nonconductor). The *rods*, R-1, R-2, etc., are metal; so are the *collector plates*, C-1 and C-2. Now suppose that the collectors are already somewhat charged with opposite charges. Then the plus and minus signs I have put on them at least roughly indicate how the charges are *distributed* on their surfaces. (You ask: Where did these initial charges come from? Good question. We must come back to that.)

Also, we have metal *inductors*, I-1 and I-2, each connected to its collector. Furthermore, we have a *neutral wire*, N, with conducting *brushes* on each end; and the brushes are now touching R-1 and R-4.

If the collectors and inductors were uncharged, there would be no reason for any charges to appear on the R-1, N, R-4 combination. But unlike charges

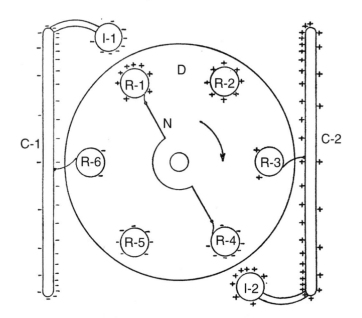

Figure 1a. Front view of a simple electrostatic generator, showing only the electrical parts.

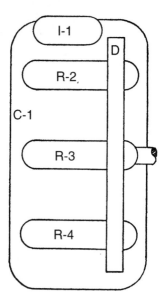

Figure 1b. Side view of the simple generator, with C-2 removed.

do attract each other. The *minus* charges on I-1 will attract *plus* charges and make them appear on rod R-1; and the *plus* charges on I-2 will attract *minus* charges and make them appear on R-4. The charges on R-1 and R-4 are induced charges. These rods are now going to become charge carriers.

Collecting the Charges

With the machine rotating clockwise as shown, R-1 and R-4 will move ahead, carrying their *captured* charges with them. R-2 and R-5 have already done this. Next, the rods (charge carriers) go on to touch the conducting brushes sticking out from the collectors. Thereupon, they leave much of their charge on the collectors. This at first may not be obvious by any means; and so, let us speak again of Michael Faraday.

Faraday showed that when you try to leave a charge on the inside surface of a closed metal can, you fail. Suppose you charge a metal ball by friction, and drop it into a deep tin can with a small mouth. The instant it touches bottom, all charge disappears from the ball and reappears on the *outside of the can*—all but an extremely tiny fraction, that is, when the can does have an open mouth. Next, use a shallower can with a wider mouth. Nearly all of the charge will appear outside, but somewhat less than with the first can. Now go to an extreme: get a metal pie pan and drop the ball on it. Even though the pan is far from completely enclosing the ball, a large fraction of the ball's charge will leave it and appear on the pan's surfaces. In our case, the charged rod R-3 does not even get very close to the collector, but yet there is enough enclosing effect to make it give up much of its charge. The shorter the brush and the closer it comes, the more charge it will give up. In Dirod I, I estimate that about three-fourths of the rod's charge is thus collected.

Generator Buildup, and Compound Interest

The number of plus and minus marks I have put on the parts of Figure 1a was merely guessed at. But if each mark did represent the same amount of charge, then every rod passing C-2 might increase the collector's total charge by, say, one-fifth; likewise for rods passing C-1.

But when the collector charges go up a fifth, the *inductor* charges also go up a fifth and the *induced* charges on rods R-1 and R-4 go up a fifth. Do you sense what will happen? As we say in engineering: "Let's put some numbers on it." Let the charge on C-2 start off at a certain amount, represented by the number 1. When it collects from the first rod, it goes to 1.2; next, 1.2×1.2, or 1.2^2, or 1.44; next, to 1.2^3, and so on. Lord Kelvin was the man who said, "It builds up according to the compound interest law." It couldn't have been said better.

In fact, a little fling at high finance will help right now. You lend a dollar to a bank at 4 percent interest, compounded annually. In one year it comes to $1.04; two years, a little over $1.08. In eighteen years, it becomes 1.04^{18}, or a little over $2.00; in thirty-six years, the bank owes you $4.10, and don't forget to collect.

Now look at those little rods in Dirod I (Plate 1 on p. 35). There are thirty-six of them. We have just found that if each one raises a collector's charge only 4 percent, one machine revolution will increase the charge by four times; two revolutions, sixteen times; three, sixty-four times, and so on. Then, if that little motor makes it whiz, think how fast the charges (and the voltage between collectors) will build up. And it really does. Do you feel an urge to build your own Dirod? I hope so!

By the way: unless something interferes, it looks as if the machine will go to millions or billions of volts. Of course, it won't. *Corona* will step in as the limiting phenomenon. We will take a close look at corona in due time.

Where Does a Dirod Get Its Initial Charge?

I love to build things. So does Charlie Hall, our instrument maker in our Electrical Engineering Department Together, some years ago, we made the parts and put the original Dirod I together. There it sat, still showing shop dirt. As a completed generator, it had never existed until that moment I turned the rotor by hand. It generated! That was one of my most exciting moments.

Now, how in the world could it generate, without my somehow giving the collectors an initial charge? First, remember that air gets ionized from cosmic rays and radioactivity. All it takes, in theory, to have a starting charge is for one collector to pick up a few extra ions of one kind or another from the air. But that is probably a minor factor in these machines.

The major factor, no doubt, is in the fact that certain parts are made of Plexiglas. These have been handled time and again, in being made and assembled. Handling, touching, rubbing—these acts put charges on the surface; and they would *induce* charges on the collectors sufficient to start the buildup.

The fact is that a Dirod will almost invariably start to generate, without help. My machines once refused to build up in Miami, Florida, where the air is salty, and the day was hot and humid. It happened once again in Maine under somewhat the same conditions. In both cases, the Plexiglas parts had not been recently cleaned. Contamination no doubt leaked away all surface charges. When this happened, I used friction to get charges established by letting the rods run briefly in contact with a limp plastic bag. There have been very few such occasions. As far as I know, these Dirods are the most rugged and reliable open-air electrostatic generators ever produced.

How Do the Charges Move Around? The Electron Cloud

If an experimenter a century ago could have looked at Figure 1, what would have been his thinking about charge, charge movement, charge density, and all that sort of thing? Well, he might have thought in terms of a mysterious something called "electric charge" that could somehow spread freely over the surface of a conductor, completely covering some part of its area, or maybe all of it, and either spreading uniformly or otherwise, depending on the situation. He had to take a lot for granted. Let us go a little way here into what really happens. We return to aluminum.

In Chapter 3, we found that aluminum atoms (in the solid form) can't leave home. But now it is time to recognize that some of the electrons can leave home. It is quite natural to think that in a neutral or uncharged piece of aluminum, each of an atom's thirteen electrons belongs to that atom and will have to stay with it. That is wrong. Most of them do. But the fact is that one or more of the electrons in the *outer shell* of an atom will go wandering and batting around in the ample spaces between atoms. Thus it is that within this little aluminum cube resting here beside me, there is an *electron cloud.* Vast numbers of electrons are constantly and forever zooming around—and on the average, going nowhere in particular. They are, however, always ready to respond to some urge, inside or outside.

Now take a new look at the induced charges on R-1 and R-4 in Figure 1. How did they get there? The plus charges on I-2 pulled some of the electron cloud electrons to the surface of R-4. *Internally*, the rods, the neutral wire N, and the conducting brushes remain neutral. Hence, on the average, the internal electron cloud moved ever so slightly toward R-4. And just as many electrons as appeared on R-4, had to leave the surface of R-1—leaving, on R-1, just as many atom protons without an equalizing electron each.

One of the triumphs of the early experimenters was to prove that when charges are induced on a conductor, the charges are not only *opposite*, but *equal.* Nowadays, it is easy to see why: every negative electron that is moved to a negatively charged area simply leaves an unequalized positive proton at the surface, in some other area.

All of which lets us make another point. When R-1 and R-4 touch the neutral brushes and the induced charges are forming, there is a brief *flow* of electrons along the neutral wire, and this *constitutes an electric current.* In fact, what we call a current in a solid conductor, such as any metal, is always really a flow of electrons. Again, however, there are so very many electrons in the electron cloud that any such flow always amounts to just a slight *drift* of the cloud along the conductor, one way or the other.

Conduction and Conductors

If all metals are like aluminum in having electron clouds within them, and if therefore they are all conductors, are they not therefore equally conductive? Not by any means. Gold, silver, copper, and aluminum are good conductors. So is sodium, when its light weight is taken into account. Some other metals are comparatively poor conductors; and many alloys are very poor in conductivity. For example, nickel and chromium make an alloy that is so resistive that wire made of it is called *resistance wire*.

If you happen to know all this, you might be inclined to use only materials that are good conductors, such as aluminum, for electrostatic generators, connecting rods, capacitor plates, or for other demonstration items. Such an idea is completely wrong. The currents to be carried in any of the electrostatic equipment described in this book are very tiny indeed. Thus, metal for conducting parts is chosen, not for conductivity, but for availability, cheapness, ease of machining or drilling or finishing, and so on.

There is one exception to that statement about these currents being so tiny. This is when a considerable amount of energy is stored in a capacitor, and the capacitor is then allowed to discharge. A large current can then flow, but it is of extremely short duration. As an extreme example, the terrific energy released when a lightning stroke occurs requires that when the stroke hits a lightning rod, the rod and its conductor to ground must be sufficiently heavy and conductive to avoid being fused or mechanically wrecked.

The Dirod Family

The plates show the Dirod family, now five in number. First came the pioneer, Dirod I (Plate 1 on p. 35), which, after numerous changes and improvements, became a thoroughly competent machine. It has been my station wagon companion for thirty-three thousand miles of lecture-demonstration driving in the past four years. Other Dirods joined in these trips as they were born, one after another.

My second generator, Dirod II (Plate 2), was designed from experience gained from Dirod I. It has fewer and larger rods, and there are some structural differences.

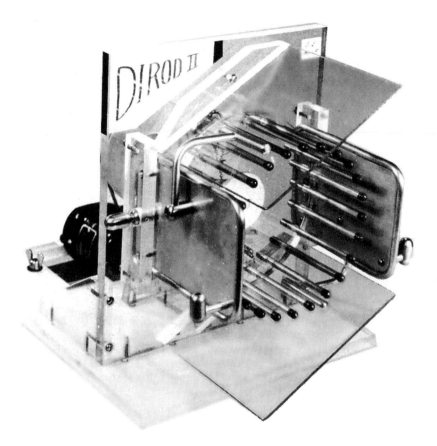

Plate 2. Dirod II. It has 24 rods.

Next came a very different design, the Radial Dirod (Plate 3), with rods sticking out from the disk like spokes in a buggy wheel. These three like to go to a top voltage of some 85,000 or 90,000 volts. And this made up the whole family when this book was started.

What brought the fourth into being was the realization that the cost and the construction problems of the first three might put them out of reach for most (but by no means all!) young scientists. Why not try to design a junior model to be within the capabilities of many persistent young experimenters? As the writing of this book proceeded, this urge became stronger. I had been dreaming of a Dirod Junior for months, and making many sketches. At last the ideas jelled, the book was laid aside for a while, and Dirod Junior (Plate 4) was born. I built it all myself, and in two and a half eight-hour working days

Plate 3. The Radial Dirod, 30 rods, with Vertical Capacitor and Sphere Gap. Radial Dirod Junior is in the background.

had it to where, turning it by hand, it generated! Another couple of days went to completing it—this included installing a motor drive. Then I went back to book-writing again.

What happened next? Another urge. This time, the editor of the Science Study Series said, in effect: "Can't you design a Dirod that is still simpler and easier to build?" Well! I thought I had done that with Dirod Junior. However, there were some ideas yet untried. With this new stimulus, book-writing was again suspended. Radial Dirod Junior was dreamed up (Plate 5); I built it all myself, and the family is now five.

These generators are all completely reliable, and any one of them will operate all of the demonstrations described later in the book. When the morning weather broadcast announces 100 percent humidity, I know that I can come to my laboratory on the campus, start any Dirod, and have it perform beautifully. I have carried my generators, uncovered, out to my car in rain, driven to a lecture, carried them in through the rain, and had them promptly swing into action. They are reliable!

In order not to interrupt our development of electrostatic phenomena, demonstrations, and applications, the description of the Dirod family will be put in Chapter 16. In the Dirod Family Data Table you will find performance

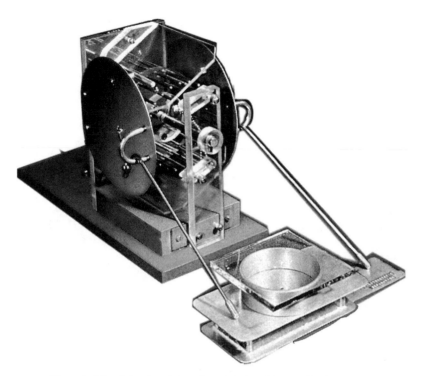

Plate 4. Dirod Junior, 24 rods, with the Vertical Ball Dance.

figures and many specifications. Also, instructions for building both Juniors are given there in detail.

Many high schools already have a Wimshurst. I have tried a Wimshurst, and know that it will operate the demonstrations to be described later. I would suggest that you improve it by giving it a motor drive with speed control

My Dirods are thus equipped. However, any of them can be rigged for hand cranking instead of motor drive.

The elementary Dirod discussed in this chapter now needs improving. We are about to do that—and learn more about electrostatics in the process.

Note. At this writing, only two of the Radial Dirod Junior generators exist. The second one is proof that a youngster can produce this machine. A grandson, A. D. Moore III, age thirteen, of Salinas, California, built it entirely by himself.

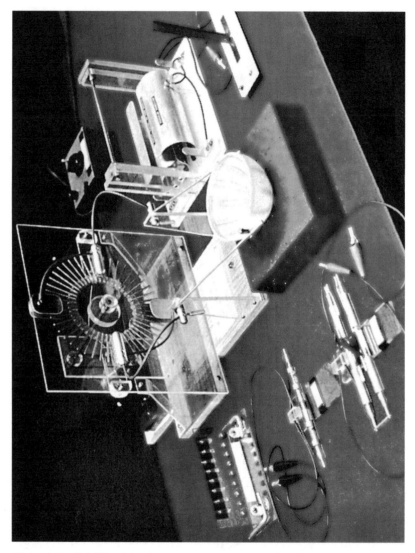

Plate 5. Radial Dirod Junior, 48 rods, with the Interdigital Motor. Shake-sphere Generators in foreground; Lamp Bank at left; Flapper and Swing Generator at Right.

Every one of these Dirods should be considered as being still under development. It takes time, thought, and experimentation to try the various changes that may lead to simplification, or improvement, or both.

5. Improving the Generator— Learning More about Electrostatics

What is the best way to learn about electrostatics? Fortunately, this is an area of science in which you can start anywhere, learning as you go along. One of the best ways is to take an electrostatic device such as a generator, and follow it through from the mere-idea stage to where it becomes a practical machine. That is precisely what we are doing here. As we develop our Dirod, we will find that it is concerned with a surprising number of the most important principles and phenomena of electrostatics, and that several lessons we learn from it can be directly applied to many other electrostatic situations.

My first generator of any kind was a Kelvin water-dropping generator, and I'll show you later how to make one. My first *rotating* generator was Dirod I. In my tours and on other lecture trips, covering thirty-five or forty colleges and universities and other places, I have told the thousands who have seen it that I have learned far more from it than it ever learned from me.

For the fact is that Dirod I, at birth, *was* little more than a mere idea. Yes, it generated—but only about 8000 or 10,000 volts. Months of intensive experimentation followed, in which many changes were made. Gradually, it was brought up to where it now has a top measurable voltage of 85,000. Thus we grew up together: it gained in performance while I gained in knowledge. I am much indebted to Dirod I. Likewise, we will now develop our main subject by taking the elementary Dirod in Figure 1, and making it better.

More Rods and Faster Build-up

The Dirod of Figure 1 has only six rods. Even so, if we added the proper mechanical parts to hold it together and make it run, it would certainly generate. In fact, if we gave it only *two* opposite rods it would generate and then be much like the early doublers. In Figure 2 we show a major improvement. I have drawn a machine with twenty-four rods, instead of two or six. If each rod carries a certain charge, then obviously, the more rods we have, the faster the charge will build up on the collector plates. Let me tell you how I tested for this when we built Dirod II (Plate 2 on p. 41). It has twenty-four rods. I first put in one opposite pair of rods; ran the machine at a certain speed; connected it to a capacitor and a *sphere gap* (to be described later); and counted the sparks per minute at the sphere gap. Then I added a second pair of rods and repeated the test, and so on, until all twenty-four rods were installed in their holes in the disk. The sparks per minute came out to be beautifully propor-

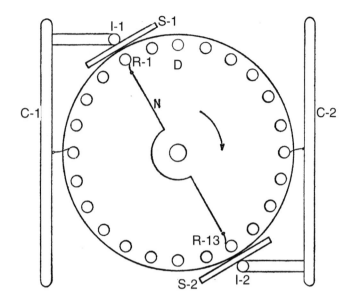

Figure 2a. Front view of the generator, improved, with more rods, and spark shields.

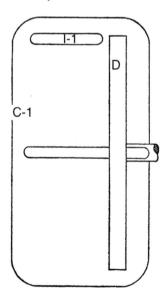

Figure 2b. Side view of the improved generator, with C-2 and S-2 removed.

tional to the number of rods. For example, sixteen rods gave twice the spark-ing rate as did eight rods. Isn't it often true that when we get onto a good thing, we make the mistake of overdoing it and come out worse? That *might* happen here. If twenty-four rods are good, why not drill more holes and put in forty-eight? After all, there is room for them. But first, that might weaken the disk so much that the rods might break out of the rim of the disk, causing an acci-dent. Second, it would crowd the rods so much, electrically, that I would not guarantee the result. For all I know, forty-eight rods *might* double the build-up rate. It would have to be tried, to find out.

We have all heard the words *voltage* and *volts*, whether we understand them or not. If our house circuits supply 115 volts, we use lamps rated for about that voltage, or else they will burn too dim or too bright. So, what *is* voltage? Well, it is the same as *difference of potential*—but then, what is that? One description of it is that it is an electrical urge: an urge to make charges move. Return to Figure 1; run the machine to charge it; then stop it. Imagine that we slide four of the rods out of their holes, leaving only R-2 and R-5. Imagine also that the bearings are nearly frictionless, to let the disk turn with extreme ease. What would happen? Why, the plus charges on R-2 would be attracted by the minus charge on C-1, and repelled by the plus charges on C-2; also, the minus charges on R-5 would be repelled by C-1 and attracted by C-2. *Result*: the rotor would turn somewhat backward before coming to rest with R-2 about where R-6 is, and R-5 coming to where R-3 is. (Or, depending on spacings, R-2 might move only a bit past where R-1 is.)

Now put the generator back together again, run it, and restore the charges shown in Figure 1. Consider the charges on R-1 and R-4. When we turn the rotor clockwise, R-1 will *resist* the turning force: those charges are being pulled against their attraction for the left-hand collector charges, and being pushed against the repulsion of the right-hand collector charges. R-4 likewise resists the turning force. *We are doing work to make these charges move.* We are putting *energy* into the system. And you can't wipe out energy: it will be around somewhere, in some form or other.

Next, arrange to turn the generator in a different way. We wrap a string many times around the shaft. We have a weight resting on the floor. We lift the weight two or three feet to tie it onto the end of the string. In lifting it, we expended muscular energy. This energy is now stored in the weight-earth-gravity system as *potential* energy. It has the potential, or capability, of deliv-ering back the energy we gave it. We release the weight, it pulls on the string, the generator runs until the weight touches the floor, then the generator runs down and stops. The energy has now appeared somewhere else. Where? Thinking of an ideal, or perfect device, with no losses, *all the potential energy first stored in the weight must now be stored by the charges we have separat-*

ed and made to appear on the collector plates and other parts. It is now stored as electrical potential energy.

This leads to the concept that there is an electrical *potential difference* between the two sides of the machine, measured in the units we call *volts*. The potential difference, or voltage between the plates, will be proportional to the charge on the plates: the greater the charge, the greater the voltage.

Voltage is often described as electrical *pressure*: the pressure that forces a current through the resistance of a lamp filament, for example. Or it may, if high enough, cause breakdown of the air, and result in the kind of a discharge called a spark. Or if not a spark, it may cause discharge by way of setting up corona, which we have yet to discuss.

It is time now to bring in an abbreviation; 10,000 volts can be written as 10 *kilovolts*, "kilo-" meaning thousand. Then we shorten that to 10 kV, and arrive at the usual way of expressing high voltage.

The measurement of high voltage can be done by using the sphere gap, in which a spark jumps between two polished metal spheres. The sphere gap, and also a *rod gap*, will be described in due course.

Generator Spark-over: the Spark Shield

Remembering Lord Kelvin's compound interest law, we know that the Dirod in Figure 1 would rapidly build up its voltage and keep on doing so until something happens. It would *spark over*, internally. A spark would jump between I-1 and R-1, with another spark, at the same instant, jumping between I-2 and R-4. Nearly all of the separated plus and minus charges would thus be joined and neutralize each other, and nearly all charge would disappear. The machine would then start to build up again.

If we want the voltage to go higher before spark-over takes place, we could move the inductors, I-1 and I-2, farther away. But this is a most unwelcome idea; for the farther away they are, the less charge they would induce on the rods, and the slower would be the build-up rate.

In Figure 2, the inductors have been moved *in*, and the rods *out*, to increase the induced charges. This shortens the gaps, and would reduce the spark-over voltage. But by installing *spark shields* as you see, spark-over is either eliminated or else could occur only at a much higher voltage. The shields are glass or Plexiglas sheets. They need to be thick enough so that they will not fail by being punctured, and need to extend far enough all around to prevent the spark from going around a shield and getting there anyway.

If this Dirod has a 5-inch disk, with 1/4-inch rods, it might spark over at around 20 or 30 kV without shields. With shields just big enough to prevent spark-over, it might go to 60 or 70 kV. Why not higher? Just operate it in a perfectly dark room, and watch: we would see a lovely display of *corona*, now limiting the generator's top voltage. So: the machine needs further improvement. After we study corona, we will enlarge these shields and then call them *corona shields*.

The Brush Problem

This may surprise you: of all the problems I faced in improving Dirod I, the brush problem was the toughest to lick. I am going to take time now to tell you all about it.

Having nothing better, I started out with little flat metal springs. They made a lot of noise, but they worked—for a while. I won't condemn this idea, for if you build a Dirod, you might do this very thing until you get better brushes. The springs worked, but they wore out pretty fast.

Next, I used fine brass wires made into a little tuft. These I got from a suede brush: a brush you can buy for brushing the nap on suede shoes. These brushes worked very well, but the tiny wires soon broke off. After all, think of the beating they had to take. If Dirod I, with thirty-six rods, ran at 600 rpm (revolutions per minute), the rods were hitting a brush with 360 impacts per second! (Later, while writing this book, I came back to this type of brush just to see if it could be improved enough so that you could use it. It could, simply by using a longer tuft. One set of brushes ran for two hours and was still in fine condition.)

Then a friend—Harold Early, a colleague here, who does research in high-voltage applications—came to the rescue. He gave me a sheet of *conducting rubber*. Now, rubber rates as a very good insulating material; but if carbon is mixed into it, it can be made somewhat conductive. From this sheet, about 1/6-inch thick, I cut little brushes. They served me very well for a few years, even though there was more wear and more friction than I liked. And they served others, for I would send pieces of that sheet to those building Dirods from plans I furnished.

In the meantime, I tried to locate the firm that had made that sheet. Harold Early could not remember where he got it. I had no luck. And then, luck broke in my favor. I gave demonstrations for a research group in The Detroit Edison Company and mentioned the brush problem. Then and there, a research engineer (and former student of mine) handed me the answer: some *semiconducting butyl rubber* in tape form. This stuff makes wonderful brushes. The friction is low, the brushes do not break, and they wear and wear and wear!

As an experimenter, I have had a rather long and varied experience in dreaming up new things and making them come true. And I learned what many another man has learned: often, the troubles you *anticipate* may be easy to surmount; whereas something to which you hardly gave a thought in the beginning may prove to be your really tough problem. The experimenter must have plenty of patience and perseverance—patience to carry a project through, even if it takes a year when the optimistic hope was to complete it in a week, and perseverance of the kind that makes one stubborn when a difficulty shows up. Instead of quitting, one accepts the challenge and refuses to be licked.

What Comes Next?

There is still some improving to be done on the elementary generator in Figure 2, but that involves corona. But we cannot properly take that up, to see why it forms and what it does, until we consider the *electric field*.

Let me remind you that up to now we have considered only *charges*: charges on surfaces; like charges repelling each other; unlike charges attracting each other; and so on. And this is much the way it was before Michael Faraday came along. Faraday began to turn his keen mind loose on something more than just charges, and to wonder about what there might be in the space surrounding these charges. He made a great contribution by giving us the concept of an electric field in the space. Therefore, a chapter on that comes next. It prepares the way to discuss corona and its fascinating doings.

6. Electric Fields

Of course, you are thoroughly aware of the earth's magnetic field. You know that if an elongate piece of hard steel is properly magnetized, and pivoted, it will be a compass: it will swing into line with the earth's field, and point to the magnetic pole. Did you also know that a long slim wire of soft iron, *not* magnetized, will also swing into line? Of course, it won't know which end of itself to point north.

Although everyone is aware of the magnetic field of the earth, and familiar with magnetic fields of magnets, hardly anyone is equally familiar with electric fields, even though they are all about us. A confusion growing out of familiarity with magnetic fields often shows up when I talk with high school students after demonstrating electrostatic phenomena to them. A student viewing a repeat demonstration may ask about the *magnetic* effects in it, while in fact it was concerned only with electrostatics. Thus, there is confusion to be cleared up; and the way to do that is to become thoroughly familiar with electric fields.

Prior to the work of the world's greatest experimentalist, Michael Faraday, there were various ideas, and much argument, about electric and magnetic phenomena. Faraday was the man whose concept of *lines of force* not only enabled him to think of, and make, important experimental advances; it stimulated the great James Clerk Maxwell to make his all-important mathematical formulations of electromagnetic phenomena, known everywhere as Maxwell's equations.

The Field of Two Parallel Rods

In Figure 3 you are looking lengthwise along long parallel rods R-1 and R-2, these having equal plus and minus charges. Along with Faraday, we now think of the charges as setting up an *electric field* in the space surrounding the rods. Some of the *lines* of the field are shown as solid lines, with arrows on them, going from plus R-1 to minus R-2.

These are called *lines of force*, as well as *field lines*, because a charged particle, such as an *ion*, would be urged to move along any line on which it found itself. Whether it *would* move precisely along the line would depend on its inertia, any velocity it already has, the effect of other particles, and so on. A positively charged particle would tend to move with the field, the way the arrows go. A negatively charged particle would tend to move the other way, from R-2 to R-1.

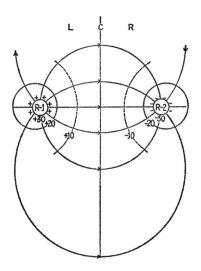

Figure 3. Electrical field of parallel rods oppositely charged. Lines of force, solid. Equipotential lines, dashed.

Let us at once take advantage of this property of the field. In Figure 3, at C, let a thin stream of small particles be released so that they drop down through the center of the field. Suppose they are of two kinds, and we want to separate them. If they have been shaken along the surface of a sloping table, and if they and the table surface have favorable electrical properties, friction effect could charge one kind positively and the other negatively. Then, if we adjust the *field strength* to the right degree, the particles, in falling through, would be *deflected*. The positive particles would fall to the right, and the negative particles to the left. This would be one form of *electrostatic separation*. More about that in due course.

As to the *field shape*, or *conformation*, take another look at Figure 3. Notice that all the lines of force, and all of the dash lines crossing them (*the equipotentials*), look like parts of circles? Well, they are, for this case. This is one of the "classic" cases, always included in studying the subject of electricity and magnetism.

The Nonuniform Field

This field in Figure 3 is a *nonuniform* field, with the field lines concentrated at the electrodes and spreading apart as you go outward in any direction. We have just considered what charged particles might do in such a field.

But what would *uncharged particles* do? Would they ignore the field? Certainly not. A nonuniform field acts on such particles with a force tending to make them *go to the more intense field*. And this is true whether the particles are *conducting* or *nonconducting*.

We again spill some particles, uncharged, at L or R, down through this field, about halfway between the vertical center plane and either rod. The field would tend to pull them to the rod. If sufficiently tiny and light, they might all be attracted to that electrode and stick there—at least momentarily. If larger and heavier, they might spill on through, but be deflected in that direction. Here again, electrostatic separation of two kinds of material could be demonstrated. It might be a mixture of lightweight sand grains and heavier metallic ore grains. The sand would be deflected more than the ore, and effective separation might be achieved.

You could demonstrate this separation effect by rigging up two jets of water to be squirted down through the field—one making smaller drops, and the other, larger drops. If the velocities are not too different, you would see an effective separation.

Perhaps by now you have thought of an interesting possibility. What if a positively charged particle falls through the left half of the field? The field acting on its charge gives a force to the right. The force toward the more intense field (which is still present) gives a force to the left. When conditions are just right, these two opposite forces might, at some point, exactly balance.

A particle in a field might have been charged by having touched an electrode. Or, as we shall see when we come to corona and ion production, *particles can be charged while out in the field*.

Let us again consider particles that are not charged. One more difference between kinds of particles is that in some materials the molecules are *polarized*, and in others they can be polarized by just being in the field; and this will affect the force acting on them. If a neutral molecule has its electrons symmetrically arranged around the nucleus, it has no external effect, and is unpolarized. But some molecules have a lack of such symmetry: the net electron position is somewhat off-center from the nucleus. It then has an external effect and is said to be polarized. Still other molecules are unpolarized normally; but when placed in an electric field become polarized when the field slightly shifts the electron arrangements from being symmetrical about the nucleus. Until quite recently, these forces on such particles in nonuniform fields have had little attention. Herbert A. Pohl and others have shown that these effects, while weak, may turn out to be potent means of separating mixtures in certain cases—not only for particles in air, but also particles in liquids. His article in *Scientific American* (see Bibliography) not only covers that, but also shows some very interesting experiments in making liquids do strange things. Another of his papers describes a fascinating hanging-drop experiment, and one in which a liquid is pumped or sprayed upward. You may wish to try these.

Summing it up, the forces on particles in a nonuniform field and the particles' responses to those forces can be affected by particle size, shape, and weight; by whether they are charged or uncharged; by their dielectric constants; and whether they are, or can become, polarized. In research and in industry there are many mixtures calling for effective means of separation; and with all of these properties at hand, there is little doubt that many new and useful separation techniques will be worked out. At the present time (as will be covered later) nearly all particle separations are achieved with the aid of corona-produced ionization, in which ions gather on particles and charge them.

Potential Difference and Field Intensity

Suppose we create the field of Figure 3 in the following way. We connect the neutral of a Dirod to *ground* by connecting it to a water pipe. (Note: no such connection is needed in the ordinary use of a Dirod.) We generate 60 kV, and connect the parallel rods of Figure 3 to the Dirod terminals. Then, since the neutral is at *zero*, or ground potential, one side has to be at 30 kV *above* ground in potential, and the other side *below* ground by 30 kV. It is the plus side, at the left, that is above ground. How is the voltage, or potential difference, distributed in space? This is shown by the several dash lines. Going to the right, from rod to rod, we read these potentials: +30, +20, +10, 0, −10, −20, −30. There is a 10 kV voltage, or potential difference, over every step I have chosen to show here.

These are equipotential lines we are talking about. *The potential along any such line is constant.* And note this: they are all *normal to* (that is, at right angles to) the lines of force where they cross. (Being thus at right angles, is expressed by the word *orthogonal*; and students of fields call the two sets of lines an *orthogonal system*.)

One way to grasp the significance of these equal potential steps is to think of moving a positive charge against the field force, from right to left: it will require of you the *same amount of work*, or *energy*, to move it across any one of these equal potential steps.

An extremely important concept, *field intensity*, comes next. It is usually stated in terms of volts per meter, volts per centimeter, or volts per inch. Take note of the short distance from +30 to +20, which has 10 kV on it; then the longer distance from +20 to +10, also a 10 kV step; and the still longer distance from +10 to zero, likewise a 10 kV step. The *volts per centimeter* will obviously be much higher in the *more intense field* near the electrode than out in the middle where the field is weakest. This will be of highest importance when we come to consider the breakdown of air, and formation of corona.

Above, we said that where field lines and equipotentials cross each other, they do so at right angles. One more thing about those field lines: everywhere that they end on the *surface of a conductor* (the electrode) *they end at right angles to that surface*. The *surface* of the *conductor* is itself an *equipotential*. Thus, when you touch a terminal of an electrostatic generator, you, being conductive, have your surface turned into an equipotential surface.

Electric Flux

Your line of thinking about the electric field must now be completed. To round out the theory, it is necessary to agree that there is an *electric flux* present in the field. Those field lines, or lines of force between electrodes, now acquire two meanings. First, as before, a tiny charge placed on any such line would have a force acting on it to send it along that line. But second, it is a *line of flux*, or *flux line* of the field. To help you get this *flux* concept, think of a little area on one electrode, having a certain amount of positive charge. From that area comes a *tube of electric flux* that reaches all the way over to the negative electrode and ends there... Well, on what? It always ends on the *same amount of charge* from which it started. But does it end on the same amount of *area*? Only in simple, symmetrical cases. In irregular fields, the ending area may be different—even vastly different—from the starting area. That is, if the electrodes are of different sizes or shapes, or both, their total charges may be the same, but the charges will be very differently distributed.

To be complete, and to avoid misleading anyone, I had better say right here that cases can occur where the flux from one electrode may not all end on the other one. (Note: this paragraph will mean more to you after we discuss space charge.) This is when the field space itself has in it extra charges in the form of ions, or charged particles, or both. Let's say they are negative ions. They *are* charges, and they refuse to be ignored. Then some of the flux from the positive electrode would end on these ions, and the remainder would go to the negative electrode.

Those ions out in the field (or other charged particles) constitute what we call a *space charge*. Space charges can indeed be important.

More about Forces: the Coulomb Force

Into the electric field in Figure 3 let us place a small particle having a positive charge. There will be a force on it. There are *two* ways of looking at, or "explaining," this force. One way is simply to say that this positive charge is repelled by the positive charges on R-1 and attracted by the negative charges on R-2.

But the *other* way is to think that the electric field somehow acts on the charge to try to move it along field lines. Now, it turns out that the actual force is truly proportional to the *product* of the *electric field intensity* and the *amount of the charge*. Very often, it is most convenient to find the force in this second manner; and the force so found is called the *Coulomb force*. In later pages, it will appear a number of times to explain phenomena and to move charged particles to where we want them to go.

Before leaving Figure 3, let us ask why the oppositely charged rods attract each other. The simple answer is just to say that opposite charges attract each other, and that's that. But now that we have these field concepts, there is another way of thinking about it. This is that first, the charges set up the electric field, with its field lines. Second, we think of the field lines as being in *tension*, and as exerting a pulling force on the surfaces where they end: that is, on the electrode surfaces. Third, it is the combined effect of all these little forces that acts to pull R-1 to the right and R-2 to the left. (Scholars differ in the way they like to interpret these phenomena. I am sure that some would prefer not to interpret these matters like this, but I find it very convenient to do so.)

Now pass to Figure 4, where parallel rods have equal but *like* charges.

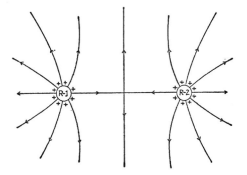

Figure 4. Electrical field of parallel rods with like charges.

They repel each other. Why? Well, like charges *do* repel! But, using the field tension idea, we see that the field has now taken an entirely different conformation. If we think of the lines as pulling on the electrode surfaces, we see that the net effect is to pull the electrodes apart.

Alignment with the Field: Field Shape Indicator

Another property of the electric field is that it tries to make objects placed in it line up with the field lines, provided they are *elongate*; that is, provided, roughly speaking, that they are longer than they are wide. You know that iron filings line up with a magnetic field and portray its shape. A similar effect occurs here. To see how I demonstrate this, go to Figure 5. In (a) we again see

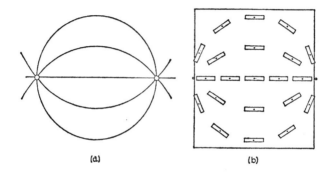

(a)   (b)

Figure 5. Field shape indicators. The little paper or card strips line up with the field.

the field of parallel rods, oppositely charged, as in Figure 3. In (b) there is a vertical panel about 9 inches wide; and the rods are seen as small circles, one at each side. The panel is insulation of some sort, such as a block of spongy plastic. Pins are stuck into the panel and placed about as shown. Mounted on each pin is a strip of paper, or else a strip cut from a 3- by 5-inch filing card, and arranged to turn easily on the pin. (Details are given in Chapter 16.) With no field on, the strips are helter-skelter. But with the rods charged from any Dirod or a Wimshurst, they align themselves as shown, and portray the field conformation.

When you dream up a new demonstration, you never know how it will "take." Will it be just one more gadget to look at, or will it be a favorite? I was surprised to find the strong and universal appeal this has for everyone—and this especially includes highly trained scientists and engineers who openly express their pleasure in it. I certainly hope you will build your own.

These little strips become *polarized* and want to behave as "electric compasses." If we reverse the field, they will *want* to reverse. When this really happens, it makes a fine show. Chapter 16 goes into this a bit further.

A Nonuniform Field and the Uniform Field

In Figure 6a we have retained a round rod for the electrode at the right, but have used a rod of odd-shaped cross section at the left. A main point to

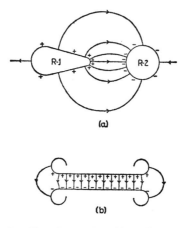

(a)

(b)

Figure 6. (a) Field of unlike electrodes. Note the concentration at sharp edge of R-1. (b) Uniform field between parallel plates.

note is the high concentration of lines on the inner thin edge of R-1, and that the smaller the radius we use on this rounding (that is, the sharper we make it), the higher this concentration will be. That is, *the sharper we make such corners, the higher will be the field intensity there.* If the breakdown of air is to occur anywhere, that is where it will take place—leading either to corona or even to spark-over between electrodes.

In contrast, in Figure 6b, our electrodes are two parallel plates. Anywhere between these plates except near the edges, we have set up a *uniform* field. The field lines are parallel. The field intensity, or volts per centimeter, is of constant value throughout the uniform field.

If two simple flat plates were used, the *edges* would, as discussed above, have high concentration of lines, high field intensity, and a high tendency to break down the air there. Therefore, we have smoothly curled the edges back all around, making field intensity weaker anywhere at the edges. This makes the intensity highest within the uniform field.

This is the way the *breakdown strength* of air is measured. It is found to be about 30 kV per centimeter for normal conditions of pressure and humidity. In a uniform field between smooth plates, the air retains its insulating value. At or above that, spark-over occurs.

The Sphere Gap and the Rod Gap

For measuring high voltage, the sphere gap (Plate 3 on p. 42) is a device that causes the breakdown of air in a reliable way, and thus can be calibrated to give quite close measurements. The main parts of my sphere gap are shown, one-sixth size, in Figure 7. These are hollow stainless steel spheres

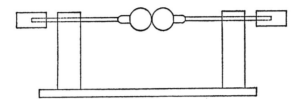

Figure 7. The sphere gap, for measuring high voltage.

1-1/2 inches in diameter. I turned the brass pieces behind them on a lathe, giving their faces conical recesses to make them fit to the spheres. They are attached by epoxy glue, with a little silver paint conductively bridging over the epoxy. (Epoxy is an insulator.) The brass pieces screw onto 1/4-inch rods. Plexiglas handles are fitted onto the outer ends of the rods. Note that the spheres are two or three diameters away from structures that would distort the field between them and cause inaccuracy.

Theory alone cannot predict the relationship of gap length to measured voltage. For each diameter of sphere used, the gap must be calibrated by the use of some instrument, such as an electrostatic voltmeter—unless, of course, the calibration has already been done and is on record.

The sphere gap is reliable because the spark smashes over *without* the preliminary formation of corona. It does, that is, within its limits of proper use: that the gap be no greater than sphere diameter. Good spheres are not easy to come by. Solid spheres are heavy, and hollow spheres are quite expensive. Turning a good sphere on a lathe has always been considered to be quite a tricky job. *Scientific American*, January, 1967, page 125, describes a method new to me. I want to try it sometime.

A *rod gap* can quite well replace the sphere gap. My rod gap is made of two aluminum rods lined up with each other, each 3/4 inch in diameter and 4 inches long.

Both ends are turned round on a lathe. The gap ends must be quite closely hemispherical and highly polished. The rod gap should give fairly accurate measurements up to 40 or 50 kV, and quite informative values from there to 60 kV.

Sphere and rod gap calibration data are given below in Table 1. In Chapter 16, I will give a table of sphere gap data for a range of sphere diameters.

Table 1. Spark gap data

kV	Rod gap, inches	Sphere gap, inches
5	0.056	
10	0.112	
15	0.170	
20	0.228	0.234
25	0.288	
30	0.355	0.370
35	0.430	
40	0.520	0.531
45	0.625	
50	0.750	0.710
55	0.903	
60	1.122	0.950
70		1.102
80		1.380
90		1.70

Induction of Charge, and Field Shape

The field shape, or conformation, changes whenever electrodes are moved or reshaped, or when objects such as conductors or insulators are placed in the field. In Figure 8, rods R-1 and R-2 are exactly as they were in Figure 3. However, a third rod, A, hung on insulating threads, has been sus-

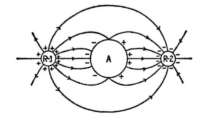

Figure 8. Charges induced on a rod placed in an electrical field.

pended in the field. The field shape changes to conform to the new situation. Charges are now induced on the new conductor. These new charges, in turn, cause some *redistribution* of the charges on the two rods, and the entire field changes to conform to the new situation.

One of the effects is to increase the field intensity near the inner surfaces of the two electrodes, thus increasing the chances for breakdown of air.

To illustrate this in dramatic fashion, suppose that Figure 3 represents a laboratory test of two power line conductors spaced a few feet apart. High voltage is applied, but there is no danger whatever that a spark will jump between them. However, a man walks in halfway between, being careful not to get close to either line. He is a conductor. He is represented by conductor A in Figure 8. A spark jumps from one line to him and from him to the other line, and he is killed. Now, since any human being can be absentminded—even highly trained men—no such test is ever conducted without being surrounded by proper precautions, and with everyone removed from the test situation before the switch is closed. This was not always so. Just about half a century ago, fresh out of college, I went to Westinghouse for a year. One day I was walking in the transformer aisle and was about to walk between two tall transformers to where there were several men. One of them bolted a few feet at me, put out his hands, and violently pushed on my chest to throw me back and away. They were about to make a high voltage test—but without proper precautions!

Conductors and Insulators in a Field

In Figure 9, two large parallel plates have set up a uniform field between them. Halfway between, we have suspended two long round rods. The upper rod is a conductor, and it modifies the field much as Rod A did in Figure 8. The lower rod is an insulator, such as Plexiglas, sealing wax, paraffin, glass, etc. It also modifies the field, gathering the field lines into itself. Note that these more concentrated lines *go right through the insulator.* This is not true of the conductor, where the lines end on the induced charges on the conductor surface.

Would these uncharged objects tend to move either way? No. They are pulled on equally each way, and would remain where they are.

You might think we could succeed in getting movement by this strategy: remove R-1 from Figure 6, remove all charge from it, and suspend it in the uniform field, pointing to the right. We hope that the higher concentration of lines on the rounded edge would dominate, and make it move to the right. Not so. If this were so, then we could resort to analogy and build a miracle space ship. We would make it of soft, unmagnetized iron, shape it to be pointed on

Figure 9. These rods, placed in an otherwise uniform field, gather the field lines into them. Above, a conducting rod. Below, a rod of insulation (a dielectric).

the front end, put it out beyond most of our atmosphere in a highly uniform part of the earth's magnetic field, and turn it loose—with the hope that without any power supply, the sharp end's pull would dominate, and the ship would move along the magnetic field lines. How wonderful! But how ridiculous!

Before leaving Figure 8, note that if we spill *uncharged* particles down through a *uniform* field, they would go right on down, not deflecting either way. To deflect uncharged particles, we would need a nonuniform field, pulling them to where the field is more intense. Of course, *charged* particles would be deflected in the uniform field.

Faraday's Famous Ice Pail Experiment

Michael Faraday, a most ingenious and persistent experimenter, had to make do with what he had in those days, or else find ways of making things to suit his needs. When he wanted to show that a charge placed inside of a nearly closed vessel would promptly disappear from the inner surface and reappear outside, he reached for an ice pail to use as the vessel.

In Figure 10 I have refined his experiment by drawing a hollow metal sphere with an opening that is closed perfectly by a metal lid. A metal ball hangs from the lid by an insulating string. Instead of cluttering up the impor-

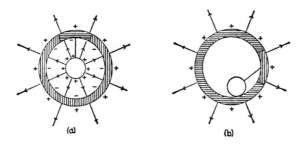

Figure 10. A Faraday cage. (a) Charged ball inserted, without touching. (b) After touching the cage.

tant issues by talking here of insulating tongs and such, just imagine we can order things around by magic.

So first there is no charge anywhere. Next we lift the lid and put a positive charge on the ball (say, by touching the ball to the positive terminal of a Dirod or Wimshurst). Then we lower the lid and ball carefully, so that the ball does not touch as it goes down through the hole. The lid is now in place (see Figure 10a). The ball's positive charge *induces* an equal negative charge on the inside surface of the hollow sphere, and an equal positive charge appears on the outside. The electric fields are as shown.

Now we tilt or roll the sphere on its side (Figure 10b) to let the ball touch. Almost instantly, all charge inside disappears by cancellation. On the outside, a positive charge remains, precisely equal to the ball's original charge.

Here we would achieve perfection. Faraday, with the near-enclosure of the open-top ice pail, achieved near perfection. In a Dirod, a charged rod, giving its charge to a flat collector, is in a still more open, or not completely enclosed, situation. Hence, it does not deliver *all* of its charge, but perhaps three-fourths of it, to the collector.

The Faraday Cage

One of Faraday's most dramatic demonstrations—and very likely his largest—was performed with a room within a room. In a large room he built a smaller room but one big enough to perform experiments in. The smaller room, now called a Faraday cage, was completely covered with tin foil. He proved that any electric field he might set up outside of the cage had no effect whatever on detection instruments placed inside. Likewise, fields set up inside had no effect outside.

As a matter of fact, a room covered with wire mesh instead of tin foil serves equally well. When you do get a chance to go through some research laboratory, and see a man working inside a wire-screened cage, you will know why. He isn't worried about mosquitoes. What he is measuring in there is sensitive to outside disturbances.

Consider the auto: except for windows and windshield, its metal makes it into a reasonably good Faraday cage. It shields you almost perfectly from the atmosphere's electric field. And if lightning strikes your car, you will almost certainly be protected. Likewise for the metal airplane. Planes are seldom struck by lightning, but when it does happen it is not an enjoyable experience. Seasoned pilots who have had it happen several times say that the last time is just as terrifying as the first. But even if the bolt does damage the plane a bit, the plane survives; apparently no one has ever been killed by lightning hitting a plane.

With plastics making such headway, there is a trend to build car bodies out of Fiberglas, which is an insulator. Such a car would offer little or no protection from lightning.

7. Corona

When nature sets up a high-intensity electric field and makes natural corona, strong men tremble and wish they had led better lives. The first sailor to see a "fire" glowing from the tip of the mast must have thought the end had come. Eventually it was named St. Elmo's fire and became a source of comfort rather than fear: for sailors on the Mediterranean, it symbolized the protection offered by their patron saint, St. Elmo. A friend, a Marine colonel, tells me of when, in World War II, he was flying a plane over the ocean in the Pacific area, with thirty men aboard. Suddenly he saw that one engine was afire. He pushed the panic button, warning everyone to get ready to ditch. When a land plane ditches in the ocean, survival prospects are not good. Then he saw that all four engines were afire. St. Elmo's fire! Such experiences are frightening.

Explorers, climbers, and camping parties trapped on exposed areas of high mountains have gone through terrifying experiences seeing corona discharges all around them and having their hair stand on end.

The complete range of electrical coronas takes in a very large territory, involving different gases and mixtures of gases; great ranges of temperature and pressure; variations in electrode size, shape, and surface condition; the use of various devices and techniques for observing the quantities involved; and the application of an ever-growing body of knowledge and theory to tie the phenomena together to learn what really goes on.

In this book we will confine our discussion to coronal discharges in air, and in gases at or near atmospheric pressure from which particles are precipitated.

Visible Corona

When friction-type generators first came along, and the early experimenters produced corona in the laboratory, how did they learn about it? Well, they *looked* at it and they *listened* to it. We can do likewise. You and I operate a Wimshurst or a Dirod in a darkroom. We have a couple of smooth 1/4-inch aluminum rods, a foot or two long, with smoothly rounded ends. We support them somehow, parallel and a few inches apart, and connect them to the generator terminals. We start the generator and watch. What do we see? Probably nothing, at first, for corona makes very little light. But as our eyes become adapted to the dark, more and more of a pale bluish haze or glow appears at the rod ends, and perhaps somewhat along the rods. If this is the first time you have seen corona, you will be entranced. I have seen it any number of times, and am still entranced.

After your excitement cools down a bit, you may point out to me that the two coronas are different! And they are. The positive corona is a smooth, uniform glow, whereas the negative corona tends to be brushy, with streamers here and there—and the streamers may move from place to place. Next, you remark about a hissing sound. There may be some lovely corona in the generator, which may be doing some of the hissing. But if you are able to pin down the electrode corona hissing, you will find that it all comes from the *negative* corona.

The smooth, uniform positive corona is giving a smooth, regular, quiet discharge. The jumpier negative corona, with its jumpy streamers, makes the irregular discharge that can be heard.

The polarity of the electrodes can also be determined by use of tiny neon lamps, called Ne2 lamps, described in Chapter 14.

Active and Passive Electrodes: The Breakdown of Air

In the preceding experiment, both electrodes were *active*. They both produced corona. The field intensity was high enough at each to cause the breakdown of the air. Now we make a change: we remove the positive rod and replace it with a thick plate with smooth, rounded edges. The positive rod now becomes inactive, or *passive*. The field intensity anywhere next to it is not high enough to cause breakdown and corona. The negative rod is still active. We see corona there.

Might there be things going on that we cannot see? Indeed, yes. Electrons are streaming out of the negative surface, and, by *electron attachment*, are joining oxygen molecules (one each) to form negative *ions*. Only a very tiny fraction of all the molecules are ionized, but it is an all-important fraction. And then? Why, these ions are propelled along the field lines by the Coulomb force, between the outer edges of the negative corona, and the passive electrode; and go to the passive, positive electrode, to leave their charges there. This flood of ions through the air amounts to a current—an *ionic current*.

Next, we reverse the show. We make the plate negative and the rod positive. The plate is again passive, with no corona. The rod is active, with positive corona. A flood of *positive ions* is now produced (oxygen and nitrogen molecules, each of which has *lost* an electron) in the positive glow; and these are propelled along the field lines from the glow, to the negative plate. Here again, we have an ionic current in the air.

Well then, what was going on when we had both rods, and both were active? Each corona was pouring out its kind of ions. Negative ions were following the field lines, from negative to positive electrodes; positive ions were doing likewise, but going the other way.

Several questions at once arise. If the two coronas produce equal numbers of ions, and if the two kinds manage to pass each other, would not much of the space between the coronas be neutral? It would be. In an imaginary, ideal case, every part of it would have an equal number of plus and minus charges, and the space would have no *net* charge. Actually, this ideal picture is modified in two ways. We cannot count on the two ion productions being equal. Also, since the two coronas are quite different, the paths on which the ions start out can be different; and, at the negative electrode, the paths shift around as this corona shifts around in its changing fashion.

Next, if unlike charges attract each other, why not expect a very large amount of *recombination*? For, if a negative ion does get close enough to a positive ion, and if velocities and lines of approach are suitable, they will recombine: the negative ion will hand its extra electron to the positive ion, which wants one, and the two molecules become neutral. Actually the outcome is that there is some recombination, but not nearly as much as you might expect.

The answer to that lies in two factors. First, the ionized molecules are few, compared to the great number of neutral molecules in which they find themselves. Second, there is the *thermal activity* of the molecules. Even though an oxygen molecule has become an ion, it is still batting around on an ever-changing path, like all the others—going straight for a tiny distance, having a near-collision, changing direction, and so on. Our recognition of this activity, which is called *diffusion*, leads to two statements. First, diffusion sees to it that opposite ions will, on the average, stay mixed for a while before recombining. Second, we said above that ions are moved along the field lines. We must now modify that by saying that *on the average* they are so moved. Diffusion makes them bob around in all directions on their little adventures, as they follow the field lines in a *general* way.

In reading this section, I hope you were not quite satisfied when I stated flatly that negative corona makes negative ions, and positive corona makes positive ions—for I didn't tell you anything at all about how this is managed. The details are somewhat involved, and possibly a bit hard to believe when you first meet them. These phenomena have taken a lot of good men through long stretches of experimental work, plus heavy thinking, before the answers—as far as we now have them—were worked out. I'll give you a simplified description of these events in Chapter 16.

Another thing about the breakdown of air: corona forms some *ozone*. The ordinary molecule of oxygen has two atoms. But electrical discharges are able to make oxygen molecules of *three* atoms, and that is ozone. So here is an example of *corona chemistry*, to be mentioned at the end of this book. Ozone is very active chemically, and it has a characteristic odor you learn to recognize. In sufficient concentration, it is a poison, just as too much salt can be a poison. I doubt if you will ever be able to make enough of it to run any risk.

Did you know that our existence on earth depends on the presence of some ozone in our atmosphere? Cosmic rays smashing in constantly maintain a small amount of ozone high in the atmosphere. Ultraviolet energy from the sun is almost entirely absorbed by that ozone. Enough does get through sometimes to give you sunburn. If it all got through, it would no doubt kill off nearly all land-dwelling forms of life.

Perhaps you know that recently EHV has come to mean *extra-high voltage*. In the transmission of electric power, the longer the power line, the higher must be the voltage at which it operates. Some lines are already up into the 500,000 volt range, and others are climbing toward the 750 kV mark. So now please take a look at the beautiful corona in Plate 6. The Ohio Brass Company kindly granted me permission to use this photograph, taken of an *overvoltage* test of a piece of a line. The test was run at 60 cycles per second, and at about 1,000,000 volts. Of course, a power line in normal use is never operated at a voltage high enough to make corona like this.

More about Corona in a Dirod

Go back to Figure 1, in which we have a simple Dirod, deliberately designed with large, smoothly rounded parts. Look again at the plus and minus marks on the collector plates. I have drawn them more concentrated at and near the edges than along the flat surfaces. They at least roughly show how the charges would be distributed. Now, the electric field set up by these charges, and by those on inductors and rods, is very complicated, and I have not tried to show it. But you can readily see that at the collectors, the most intense field would be at the edges. If the voltage ever went high enough, corona would certainly appear there. However, this machine has no spark shields, and it would spark over internally before the voltage gets up to where corona would be produced.

In Figure 2, with spark shields, it builds up to its *corona-limited* voltage (if nothing is connected to it that would itself limit the voltage). Corona would appear on each side of the generator somewhere; and as fast as charges are delivered to the collectors, they would be leaked away through the air by ionic currents set up by the coronas. But just *where* would the corona be? That is one of the fascinating things about this: if we build the machine as an exact copy of this drawing, no man alive is smart enough to predict the answer to that question. The reason is the extreme complexity of the electric field. It is

Plate 6. Corona—produced in a million-volt test by Ohio Brass Company, on a section of a power transmission line. This is an overvoltage test: power lines at normal voltage do not make corona such as you see here.

impossible to be sure of where the highest field intensity will be. So what do we do? We go ahead and build it, operate it in the dark, and *look* at it.

When I built Dirod I (Plate 1 on p. 35), the machine had so many sharp points and edges that it would only go to 8 or 10 kV. Therefore, what little corona it did have was too weak to see. But after being rebuilt to eliminate some weaknesses, there was some corona to look at. Much of it was at the inductors, with corona spreading along the surfaces of the spark shields and falling over the edges to the rods. Some of it came frontwise, to fall to the rods' outer ends. Much of it, if we describe it at the upper shield, climbed the shield and spilled over in a Niagara Falls to the passing charged rods. A very beautiful display! This was when the machine told me that the shields are not only spark shields, but must be *corona shields* as well, and that they were too small. When I replaced them with the larger shields you see on Dirods I and II (Plates 1 and 2), this corona leakage was reduced—whereupon the voltage went higher and corona broke out elsewhere.

As the game of corona-suppression and machine improvement went on, rod-end corona appeared in Dirod I. At the rear, corona streamed over the disk surface by way of the metal hub, between the charged rods at upper right and those at lower left. In front, a beautiful sweep of corona went from the one set of rod ends to the other. A lovely display it was, to be suppressed (with some regret!). But could it be suppressed? It could be. The sleeves you see slipped over the rod ends are pieces of Tygon tubing, each plugged at the end with epoxy glue.

As a measure of the sleeve effectiveness: if I remove the front sleeves from Dirod I, its voltage measured at the sphere gap drops from 85 to 70 kV.

In Dirod II (Plate 2 on p. 41) the reason for using larger rods (1/4-inch instead of 1/8-inch) was to reduce the high-intensity fields at the rod ends. The front ends were given a thick coat of corona dope after some corona was observed there. The rear ends get by without treatment.

Another change used in Dirod II was to reduce sharpness of collector plate edges by putting on *corona trim* around the edge. This is 1/4-inch aluminum rod, bent to go all around, and held by epoxy glue. Since Dirod I gets by very well without such trim, I doubt if the trim adds much to the corona-limited voltage of Dirod II.

Balanced Corona, and an Analogy

Think of a tall water tank, somewhat more than 100 inches high. A hose steadily pours in water. Forty inches up, there is a hole in the tank large enough to spill all of the incoming water. The water rises to 40 inches, but no higher. We plug that hole, and the water rises to 70 inches, where there is another hole. We plug that. Then it goes to 90 inches, to a third hole, and stops rising. This is a crude analogy to what we do when we build a new and different electrostatic generator (or other high-voltage device). We see a corona leakage, perhaps at 40 kV. We plug it, so to speak. At 70 kV, a new corona leak appears; we plug that; and go on up to where a third leak sets the limit.

Now start over again. In a new tank, there are three holes, *all* at 90 inches. When the water gets to 90 inches, all three holes spill water. We plug any two holes, but the third spills all of the water and stops the rise. That is, if you get your generator improved to the point where, in the dark, you see a rich display of corona at *several* locations, you may have reached close to what I call a *balanced-corona design*. If you take steps to shut off or reduce one or two coronas, the others will give you a richer display and greater leakage, but will usually not permit the machine (or device) to go higher in voltage.

Engineers with plenty of experience, and the use of theory as far as it will help, can often do a pretty good job at designing new and different high-voltage apparatus. Even so, the more the apparatus, or EHV power line, or whatever, differs from that which has gone before, the more essential it is to give it a thorough testing. To remedy the weaknesses that show up, modifications often are made.

Generator Voltage Not Fixed by Speed

You have a Wimshurst or a Dirod running at low speed, and the highest voltage you can measure at the sphere gap is, say, 70 kV. If you double the speed, will the voltage go higher? No, or not much. It is characteristic of electrostatic generators that the top voltage is not fixed by speed. This has long been known.

At the lower speed, corona leaks here and there were letting the charges drain from one side to the other as fast as the machine separated the charges and delivered them to the collectors. At the higher speed, the coronas already formed simply got richer (putting on a better display in the dark) and proceeded to leak larger ionic currents through the air at the same voltage.

Suppressing Corona

The problem of corona suppression, and the ways of solving it, have already been pretty well illustrated in the foregoing. In your own experimentation, you may work up a new high-voltage demonstration or build your own new design of an old one. If your judgment and your luck are both good, you may have it work splendidly. Or again, it may be a complete flop. Perhaps there is a sharp point or edge you have overlooked, making such a severe drain that you just can't get the voltage high enough. Maybe an innocent-looking screw, needed for purely structural purposes, is the villain: it may go completely *through* an insulating piece, when it should only go part way in and stop in a blind hole. Maybe two electrodes, or connection rods, are too close together; if separated, their high-intensity fields will be lowered (but there is a limit to this—changing from 2 inches to 6 inches may work wonders, but beyond that may do little for you). Or again, a corona block may be feasible, by insertion of a sheet of insulating material. Perhaps the end of a rod needs a knob on it. Or if corona sprouts from a limited place, application of corona dope may shut it off.

Suppose a machine screw, needed to hold some part, had to be installed so that its head is exposed on the C-2 collector plate in Figure 1. If the exposed head is at the very edge, corona might surely sprout from it. Placed *in* from the edge by half an inch or so, it is in a much less intense field, and may give no trouble. Placed an inch or so away from the edge, the field there is still weaker, and you need not worry about it.

Thus, the lessons learned from some one device, such as a Dirod, may help you out in creating a prize-winning demonstration of your own for a Science Fair.

8. The Electric Wind; Other Corona Effects; Precipitation

A corona discharge in air sets up a movement of the air called the *electric wind*. This began to be sensed two centuries ago. But no good chance came along to observe it until friction-type generators enabled experimenters to produce corona continuously. In "A History of the Electric Wind" Myron Robinson states that Tiberio Cavallo worked out the correct explanation of it in 1777. Until then, and even later, various ideas were entertained about it, for it was of great interest to many electrical experimenters.

In reading about corona phenomena and the useful applications of corona, always keep in mind the tremendous number of elementary charges flooding out from the corona surrounding an electrode. Consider a negative electrode with corona amounting to only one microampere (a millionth of an ampere). There are then 6,230,000,000,000 electrons *per second* leaving the electrode. In air, most of these soon attach to oxygen molecules, so that a torrent of about that many negative ions per second is released into the electric field space. For a positive corona of one microampere, an equal flood of positive ions is produced.

The Electric Fly, or Pinwheel

The electric fly, as it has long been known, makes a striking demonstration. One way to construct it is shown, about one-quarter full size, in Figure 11. The basic part is a wire bent in the form of a long Z and mounted on a bearing and support rod. The design I show here uses a broken piece of a needle for a shaft, with a Plexiglas block drilled to take the shaft, and the wire glued to the block. As drawn, the needle and the wire nearly touch, but not quite. If they touched, it would add to the friction. I think it will work this way. If not, replace the Plexiglas block with metal, to make sure of electrical contact with the metal rod and metal base.

If one terminal of your generator is connected to the base, the pinwheel spins—backward! More strikingly, one can hold the base with one hand while touching a generator terminal with the other. It turns. Why?

First, there is corona discharge at both points. Second, the corona is loading the air near the points with ions. Third, these ions would be propelled away to follow the field lines, if free to do so. But fourth, they are mixed among the vastly greater number of neutral air molecules, and have to drag these, more or less, along with them. Fifth, and consequently, the air near the

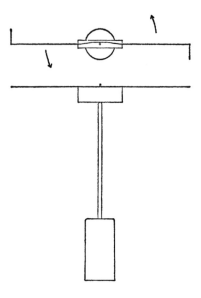

Figure 11. The electric fly, or pinwheel.

points is repelled by the points, and the points are repelled by the ions. The pinwheel turns backward.

It is still more fascinating when I set my pinwheel close to and directly in front of a Dirod. Without *any* connection, it operates. Seeing this certainly helps one to believe in ionic currents. My pinwheel is to the left of the electric blower (which we come to next) in Plate 7.

The Electric Blower

My electric blower is shown one-third full size in Figure 12. It is also seen, center front, in Plate 7. The box, open at both ends, is made of Plexiglas. All other parts are brass or aluminum. In use, connecting rods from a generator are laid on the top electrodes. When smoke is released near the left end, it is rapidly drawn through the blower.

A strong electric field is set up between the point and the grid. Corona formed around the point produces ions that are repelled to the right, and these ions drag the air in which they are mixed, along with them. Hence the electric wind, here boxed in, makes a blower. Here again we turn to Myron Robinson, who has done thorough research on the blower. My blower is much like one he describes in his paper "Movement of Air in the Electric Wind of the Corona Discharge." The electric blower offers two advantages: it is silent and it

Plate 7. The Electrostatic Zoo. (These animals will bite, but only in play. Not dangerous.)

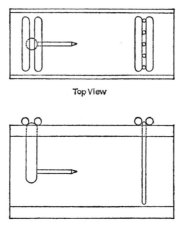

Top View

Side View

Figure 12. The electric blower.

has no moving parts. However, as Robinson has shown, it is probably inherently limited to a top efficiency of 1 percent. But as an electrostatics demonstration, it is eminently successful. All who see it are much impressed.

While writing these very lines, I stopped to run a test. The big Radial Dirod (Plate 3 on p. 42) went to 90 kV as measured at the sphere gap. Then, connected to the blower, the voltage went only to 18 kV at low speed; and even though this generator has much the largest current output of all my Dirods, running it at top speed raised the voltage to only 20 kV. This is another case of corona limited voltage. The generator's own corona limited the machine to 90 kV in the first test. In the second test, the *blower* corona limited the system to the far lower value.

This reminds me of a happy idea I had some years ago. Why not make the wind from an electrode drive a paper paddle wheel? So I made a nice paper paddle wheel and tried it. And it ran. But it ran backward! Delightful surprises like this may show up any time, when experimenting with electrostatics. Later on, it quit running. Someday I want to take time to duplicate this affair, for I have no idea why it ran that way, or why it quit. (An experimenter never has time to do all that he wants to get done!)

Every so often somebody gets the idea of using electrostatic forces, such as the electric wind, to cause levitation: why not support an airplane, and perhaps even drive it? I understand that one well-known individual did succeed, in recent years, in floating a small, very lightweight model electrostatically; but *not* by having it carry its own power supply. That remained on the floor. I never happened to read the published account of it, but it certainly made a deep impression. On my tours across the country, plenty of questions were raised about it. My answer then—and now—is that whereas it may be done with a very small, very light model not carrying its power supply, it is utterly and completely out of the question to be able to levitate a plane large enough to be of practical use.

This blower is also, to some extent, a *precipitator*. A fair amount of the smoke is trapped by deposition on the grid. After a number of demonstrations, the grid and the interior will need cleaning. I recommend that you put the box together with screws, not epoxy. Cleaning is much easier if you can take it apart.

Demonstration of Smoke Precipitation

In Figure 13 you see a glass jar with a plastic screw cap, shown one-third full size. A pointed electrode sticks down through the lid. I have also drilled

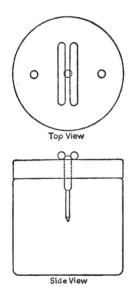

Top View

Side View

Figure 13. Smoke precipitator.

two holes through the lid. A straw stuck through a hole is used to fill the jar with smoke.

When one terminal of a generator is connected with the electrode and the generator is started up, two things happen in rapid succession. First, the smoke violently churns and swirls. Of course, corona formed around the point is making an electric wind, and it churns the smoke. Next, the smoke all disappears! If this is repeated enough times, the inside of the jar becomes so coated that it needs cleaning. Thus, we know where the smoke goes: it is electrostatically *precipitated* on the glass. But why?

First, the electrode sets up an electric field going out in all directions. Second, corona around the point makes a flood of ions, which move along the field. But third, these ions promptly collect on the smoke particles, making them *charged particles*. Fourth, these particles, propelled by the Coulomb force, are deposited on the glass. Once again we turn to Myron Robinson, this time to his paper "The Origins of Electrical Precipitation." He shows that a German teacher, M. Hohlfeld, demonstrated this in 1820. His device required a spark to do the trick; but C. F. Guitard, three decades later, showed that a corona discharge would do an effective job of precipitation.

The Uncertain Soap Bubble

You blow a soap bubble, turn it loose, and approach it with a highly charged electrode, as in Figure 14. Can you predict what it will do? I have

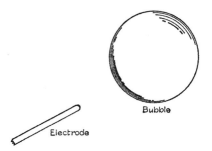

Figure 14. The uncertain soap bubble.

tried this many times and can I therefore predict its behavior? Emphatically not! This demonstration is a beautiful illustration of the fact that *things that are very simple in appearance may really be subject to a number of complex and unseen forces.*

Three purely mechanical effects are there to begin with. First, the weight of the bubble's water film tends to make it fall. Second, your breath, if warmer than room air, gives buoyancy, tending to make it rise. Third, room air is always on the move, tending to drift the bubble in some random direction. Now for electrical effects.

Your generator is running, developing high voltage. You have connected that electrode to one terminal by a flexible, insulated wire. You make the approach. The bubble, being conductive, is attracted to the electrode, tending to move into the more intense field. But corona around the end of the electrode makes the electric wind, tending to blow the bubble away. Also, ions from the corona rapidly gather on and charge the bubble, also tending to make it be repelled. If we ignore the three mechanical effects, we still have three electrically produced forces, all unknown; and we cannot know whether the bubble will snap to the electrode and be wrecked, or whether, using the electrode, you can chase it around the room. Either may happen. When the chase develops, you have a fascinating demonstration.

Now, this isn't all—not by any means. Three questions: When you *blew* the bubble, were you holding the bare electrode in one hand? Did you blow it from the end of a *metal* tube? And what kind of shoes were you wearing?

If the shoes were leather and the floor is not a good insulator, leakage through you to ground may have grounded the electrode, perhaps eliminating its corona. Shoes with rubber or plastic soles may pretty well eliminate that. Next, holding the electrode means that you were charged, and, by way of the metal tube, the bubble was already charged—an effect not covered above. So then it occurs to you to rig up an insulating handle for the electrode, so that you won't be charged. But then if you stand fairly close to a generator that is making full voltage and plenty of machine corona, you are being flooded with ions produced at its two sides. One kind or the other may predominate, and you are getting somewhat charged anyway. If you really want to blow a bubble with no charge, blow it with an insulating tube.

I have used a metal tube to blow the bubble, while holding a generator terminal, and the tube then is also the electrode. The released bubble sometimes goes through a beautiful arc, following the field lines and landing on the other side of the generator. Very pretty!

The uncertain bubble has been discussed at length here, for the very good reason that you may work up some other demonstration, and have trouble. Why doesn't it work? Or why did it work the first time, three ago, and never again? What changes did you make that may have seemed unimportant at the time? In this game of playing with unseen forces, you sometimes have to think of everything that may have an effect.

Bubbles again: if you have never read that famous and fascinating classic, *Soap Bubbles* by Sir Charles Vernon Boys, you should. He describes some mighty intriguing electrostatic effects with bubbles and water streams.

The Electric Wind and the Candle Flame

Plate 3 on p. 42 shows the Radial Dirod connected to the vertical capacitor (described later) and to the sphere gap. Disconnect the capacitor and remove the sphere gap. We then have two electrodes sticking out horizontally. With the generator turning, the coronas at the ends of the rods make electric winds, and these can be felt by holding the hand an inch or two away. Why not make the wind effect visible by having it blow a candle flame?

Hold the flame close to the positive electrode end, and it does just what you would expect: it is vigorously blown away. But, held at the negative electrode, the flame divides! When this works at its best, about half of the flame blows away, and the other half is attracted to the electrode. Clearly, something needs explaining.

This demonstrates an effect discovered long ago: *that a flame is positively ionized.* Some of the positive volume is attracted to the negative electrode; for the remainder, the wind effect predominates. As I understand it (and I hope I'm right), the process of combustion produces many electron-positive ion pairs. The electrons are of far less mass than the ionized molecules; they have far greater *mobility*; thus, many more of them escape from the flame, as compared with the far heavier ions. That leaves the flame with a net positive charge.

When I demonstrate this to an audience, I never know ahead of time about the polarity of the electrodes. So I am sneaky about it. I first rapidly pass the flame in front of the electrodes, to see which is which. *Then* I settle down with the flame at the *positive* electrode, and all of the flame blows away. Having already described the electric wind, this is just what the audience will expect. Next, shifting to the negative electrode, a good surprise is achieved.

Electrostatic Precipitation Put to Use

Is electrostatic precipitation any more than just a laboratory curiosity? By the time you finish with this chapter, you may be amazed at how it has been put to use. It has quite a long history.

When Wimshurst began bringing out his giant electrostatic generators, various workers began to get ideas about practical precipitation; but these early efforts really got nowhere. Sir Oliver Lodge, about 1905, dreamed of clearing up London's smokes and fogs electrostatically. The dream didn't come true, and we still cannot tackle that large a job in this way.

It remained for one of America's great scientific pioneers, Frederick G. Cottrell, a physical chemist, to break ground and initiate the first really successful installations of industrial electrostatic precipitators. He worked in the

University of California at Berkeley. By 1906, when he began to get results, he was able to use the electrical output of the transformer-rectifier combination to furnish high voltage and provide the needed amount of direct current—more than electrostatic generators could supply. Cement mills, ore treatment plants, smelters, and chemical plants that had been pouring smokes and dusts and poisonous fumes out of their stacks, sometimes polluting and poisoning great areas around them, began to be tamed. Cottrell precipitators trapped the solid or liquid particles. Moreover, in some processes, recovery of acids or mineral products that had been wasted, helped to pay for the process, or were even made to yield a profit.

Cottrell was both a dreamer and a doer. As a doer, he was *the* pioneer who, with great ingenuity, started electrostatic precipitation on its remarkably successful path. As a dreamer and benefactor, he organized the Research Corporation and turned some of his most valuable patents over to it, the purpose being to use the income to make grants to others needing funds for research. Two of the earliest grants made brought notable results. One was to Ernest O. Lawrence, for developing the cyclotron; the other, to Robert J. Van de Graaff, to develop his generator. By now you may suspect that Cottrell is one of my heroes. He surely is; and he will be one of yours, too, after you read Frank Cameron's absorbingly interesting book, *Cottrell, Samaritan of Science*. Every scientist and engineer should read it.

Much of the early work done by Cottrell and others was done "in the dark," for the theory was by no means developed. Hunches, experimentation, and trial and error had to help out. And of course there were failures along with successes. Gradually, as more and more was learned about corona, ionization, and other factors, precipitation processes were brought under better understanding and control. Today, nearly all new installations can be directly designed and engineered with the certainty that they will work pretty much as planned. It is interesting to note that the world's first book on *Industrial Electrostatic Precipitation*, by Harry J. White, was brought out only a few years ago.

Today, great numbers of electrostatic precipitators are in use. They range from little air purifiers for the home; through larger ones for hospitals, and "clean rooms" used in assembling delicate electronics parts in industry, and so on; up to the veritable giants that clean up the flue gas from coal-burning powerhouses. These last, incidentally, make those tall stacks seem useless, because often you can barely see anything coming out of them!

How These Precipitators Work

The usual basic unit in a precipitator consists of a small active electrode and a large passive electrode, with enough voltage between them to form corona on the active electrode. The higher the voltage, the better. Therefore, it is made as high as can be, without spark-over happening, or else with only occasional spark-over. Small household precipitators operate at a few kV. Industrial units may range from 30 to 100 kV.

The power supply furnishes unidirectional voltage. That is, it is a one-way voltage, not reversing as in AC; but typically, neither is it constant. It is alternating current as to source, but *rectified* (made unidirectional). It has somewhat of a wave form, but nevertheless it works very well. In fact, in some applications pulses are used.

It is common practice for the active electrode to be a wire, 0.1 inch or less in diameter, tightly strung along (or up and down) the center of a round duct; or to be several wires across the center of a rectangular duct; and there may be one or a great many such units.

The air, or gas, is engineered to flow along the pipe or duct at something like 10 feet per second. Floods of ions from the corona attach to the particles, which are then moved *across* the flow by the Coulomb force, along the lines of the field, to the pipe, or passive electrode. If liquid droplets are being trapped, they will of course flow down the duct walls. But when the particles are dry, such as fly ash from coal combustion, it presents problems. How to knock it down automatically and collect it without knocking it out into the air stream, has called for much ingenuity. In some processes, it is arranged to wash down precipitated dry dusts.

It is common for collection efficiency to be 95 percent; many times, it goes to 98 or 99 percent; and some cleaners are required to have 99.9 percent efficiency. At that last figure, one particle in a thousand would get by.

One of the great pioneer installations was made in 1919 at Anaconda, Montana, for the Anaconda Copper Smelting Company—a gas-cleaning operation for cleaning a flow of 2,000,000 cubic feet of gas per minute.

The pioneer installation of a *fly ash* precipitator was made in 1923 in my own state of Michigan, for The Detroit Edison Company. Today, a large fraction of electric power is produced by burning fuel. A large fraction of such powerhouses burn pulverized coal, and most of them are near enough to populated areas to require fly ash precipitation. And about how much of it is precipitated in this country? I am indebted to Myron Robinson for this estimate: about 20,000,000 tons per year!

The fly ash precipitator in Plate 8 is an Opzel Plate Precipitator, made by Research-Cottrell, Inc., of Bound Brook, New Jersey. The picture was kindly

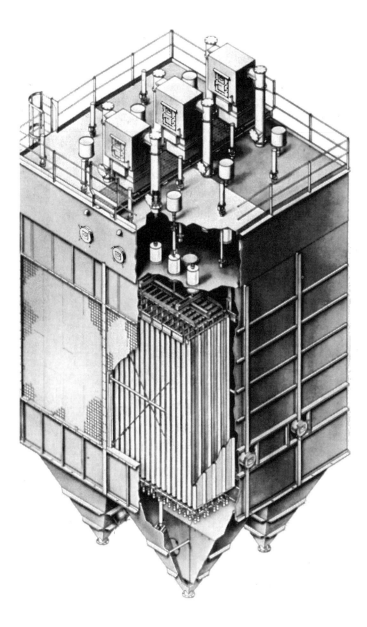

Plate 8. An electrostatic precipitator, made by Research Cottrell, Inc. Three stories high, it handles 200,000 cubic feet of gas per minute.

furnished by the company, and included here with their permission. It is 36 feet high along the edge of the rectangular box, and it handles 200,000 cubic feet per minute. Yet it is just a baby when compared with the world's largest, very recently built for the Consolidated Edison Company of New York. This giant is larger than many sixteen-story apartment buildings, and it handles over 3,000,000 cubic feet of gas per minute.

Enormous volumes of gas pouring out of blast furnaces and electric furnaces, and from many processes in the chemical industry, are cleaned up electrostatically. It should also be mentioned that the process has a number of applications to *liquids*, such as removing water from oil, or solid particles from a liquid.

The Coulomb forces moving charged particles to the collecting electrode are weak, not strong. Thus, it is confined to collecting small or lightweight particles. But within its capability it is giving superb service in many applications. And it is, of course, a major factor in that ever-growing problem, the control of air pollution. Moreover, it is inherent in the process that very little energy is needed to maintain high collection efficiency.

9. More Demonstrations; Separation of Mixtures

If you build a Dirod or have a Wimshurst, you will want to have all the fun you can with it. The fun may be of three kinds: making your demonstration items, enjoying their operating, and learning why they manage to do what they do.

In this chapter I will describe a number of demonstration items. The last one will serve to introduce another large family of applied electrostatics: separation of mixed granular solids.

The Little Island That Runs Away

For this charming little trick, your apparatus will be simple indeed. Take a glass, plastic, or china bowl or cup, and pour in water, as in Figure 15. It is

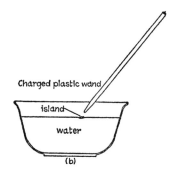

Figure 15. The little island that runs away.

shown one-third full size. Next, you float some little pieces of metal on the water. Will metal float? Certainly, if the piece is small enough, and if you slide it gently onto the water from a piece of paper. For metal, I use cast iron shot, having lots of it for other purposes. You can take nippers, and bite half a dozen little bits of metal from the end of a wire. Any little bits about a millimeter or so long and wide, will do. The bits will soon float together and make a little island, floating in its own depression, and held up by surface tension.

Now for the magic wand. This is a plastic knitting needle. Just about any other rather pointed piece of plastic will do. You now electrify (charge) the needle by drawing it through your coat sleeve or other cloth, squeezing as you pull it through the folded cloth. Then prove that it is charged, and will attract things, in three ways. Let it pick up tiny bits of paper. It does. Wipe it clean, recharge, make it pick up ashes. It does. Wipe it, recharge, and make it pick up some extra bits of metal you made and laid on the table. It does. Well then, if you wipe it and recharge, and make it approach the little island, it should attract the island, shouldn't it? Try it and see! No matter how you make the approach, the island runs away!

I discovered this thing by accident, and was certainly astonished at the time. Since then, I have puzzled many others with it. If I had only kept a record of all the foolish explanations I have had to listen to—even from highly trained men!—it would make very interesting reading.

The truth of it is simple indeed. The charge on the needle's end induces opposite charges on the water. The water is then pulled up into a little hill. As you move the needle along, the hill goes with it, and always pushes the island, in its depression, away.

At this point, you may have gone back to my drawing to look for that little hill. It is not there. I deliberately left it out. When you demonstrate for others, they will "leave it out" too; they won't see it, and only very rarely will someone think of it on his own.

Another discovery I made with this should also be tried. Take the island out, or ignore it. Dim the room lights. Charge the needle to the limit, then, with a very steady hand, make it approach the water. Watch it from the side, closely. When you have it just the right distance up, you see the hill form—and rise—and come up *almost* to touching. But then a tiny spark jumps, and down it goes. *But don't quit.* Soon, the hill forms again, the spark jumps, and down it goes. Believe it or not, with one charging, this once repeated for me thirty-five times! Thus we learn that as much of the charge is removed from the tip of the needle, some charge farther up creeps down, to repeat the phenomenon.

The Electrified Water Spray

In Figure 16 you see a plastic bulb, with a cap that has a small orifice. What came in it, I forget; but I do recall that when empty, my wife let me have it for my laboratory. When I fill it with water and squeeze, it squirts a thin stream halfway across the room. This, by itself, always intrigues any audience.

Notice that I have installed a screw in the bulb, with the nut inside. So next—wearing shoes with rubber soles—I touch one terminal of a generator, and that charges me. My thumb on the screw gives me contact with the water, and that charges it. I squeeze again. Instead of the great long stream, I get the

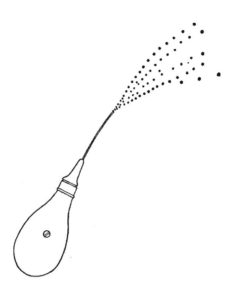

Figure 16. Electrified water spray.

effect you see in the drawing. Only a few inches away, the charged water breaks up into a lot of separate streamlets (more than I have shown) and each of these breaks into droplets. A lovely display.

You could go further. Have an orifice still smaller, to make a finer stream and smaller droplets, and have a friend standing near and touching the other terminal. You aim the droplets to go past him, but: he gets sprayed. He will be impressed. And you will have demonstrated, in a simple way, something of what goes on concerning what we come to soon: electrostatic spray painting.

Another stunt that always gets delighted attention is to have the electrodes sticking out as they did for the candle flame experiment. Place a single large drop of water on top of the knob on the end of the electrode, and start the generator. The high field intensity will draw the drop out into a fine point, from which will come a fine spray of tiny droplets.

The Clapper

My clapper, shown one-third full size in Figure 17, is a charge-carrier and a noisemaker. The two vertical side pieces are aluminum electrodes, to be connected to the generator terminals. They are attached to a Plexiglas base, and at the top to a Plexiglas block. A 1/8-inch rod sticking out from the block

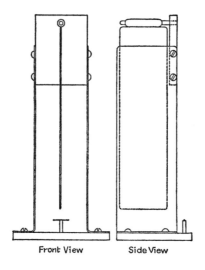

Front View   Side View

Figure 17. The noisy clapper.

supports the swinging aluminum clapper. In action, the clapper bangs back and forth, carrying charges from one electrode to the other.

If the generator speed is such that its output cannot all be used by the clapper's charge-carrying function, the voltage builds up and the device sparks over. This interrupts it, and it has to start over again. To help prevent this, I installed the short piece of wire on the little post. You see the post between electrodes, standing up from the bottom. So located, there is an intense field at each end of the wire, corona is formed, ions are made, and the ionic current keeps the voltage down. In effect, the piece of wire is a crude voltage regulator.

Franklin's Electrostatic Motor

Benjamin Franklin, who discovered and started so many things, invented the world's first electric motor. It had an insulating disk with metal balls attached, and two electrodes. And it ran. It ran either way, depending on the start given it. My motor (Figure 18) is shown one-third full size. It also appears in Plate 7 on p. 75. Its base and panel are of good insulating material. The disk is 1/8-inch Plexiglas, with conducting balls embedded in the rim. The two round electrodes are pieces of tubing I happened to have, and they stick out from the panel. Smaller electrodes would do. There is an arrangement to connect the generator to them from behind. The shaft is a broken

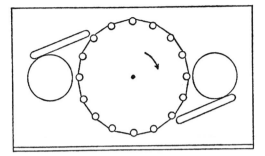

Figure 18. The Franklin motor, modified to be self-starting and one-way running.

piece of needle, supported by a base behind the disk. There is no bearing, other than a hole drilled through the disk center. A bit of oil suffices.

The drawing shows a vertical panel. I actually sloped mine back a bit. It looks better that way, and there is less chance of the disk dropping off when someone picks up the motor.

Having given a good deal of attention to the *induction* of charge, you may well guess that the motor somehow uses this effect. Not so. Operated in the dark, you see a *brush discharge* from each electrode to the passing balls. Thus, the balls are charged by *contact*. Not rubbing contact, it is true; but contact in effect, by way of those discharges. So, if the right-hand electrode is positive, it is repelling positive balls leaving it, and attracting negative balls coming to it. The same things occur, but with reverse polarities, at the other electrode. Note that the balls are serving as charge carriers, and that the electrical output of the generator is turned into mechanical power to drive the disk. My disk has run as fast as 2900 rpm.

I like to improve things, and I was seized with the idea of making the Franklin motor self-starting and one-way running. But how to do it? I was quite in the dark, and so just started to try things. There followed fascinating and frustrating experience, sticking both conducting and nonconducting materials of various shapes and sizes and locations, onto the electrodes. Sometimes, a bit of self-starting would show up—only to get balky again. Finally, I tried the auxiliary rods you see: brass rods attached to the electrodes. And they worked! I didn't know why, but they worked. Much later only, did I suspect that the electric wind from the rod ends starts the disk, and helps to run it. I *think* this is the answer.

Figure 19 shows, one-third full size, a cardboard disk with a Plexiglas hub, mounted on an insulating handle, and with brass washers epoxy-glued to the rim. If the regular motor's disk is removed and this one held in place, it operates too. Now, with the regular motor still running, suppose we hold this

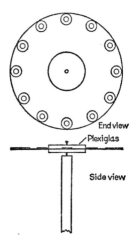

End view

Plexiglas

Side view

Figure 19. A motor disk of cardboard and brass washers.

cardboard disk close to the Plexiglas disk. Air drag should make it turn *with* the rotor. But just see what turns up when you try things. A student in Arizona, on one of my tours, tried this, and he discovered that it runs the other way! Explain that, if you will. (What happens is that the electrodes charge the washers, and these charges react with the balls' charges, and it is thus driven backward electrically, *against* the air drag.)

We can get another lesson out of this disk. To protect the cardboard disk, make it look better, and help keep it from being affected by high humidity, I coated it with Krylon—an insulating coating that comes in various colors in spray cans. For a year or two it did the running-backward stunt in fine style. Then it quit. Mystified, I laid it aside for some weeks. Then it came to me: even though it *looked* clean enough, a lot of sweaty hands had handled it. I washed it, and it again performed. Never forget: a surface may *look* clean, but contamination can be there, causing plenty of surface leakage.

To make my motor disk, I started with the disk *larger* than shown, so that in drilling the ball holes, the drill would not break out. After drilling, I sawed and sanded the disk edge down to let the balls be exposed when inserted. The balls are epoxy-glued into their sockets.

These balls are rather special: solid polystyrene, copper-coated, hence, very light in weight. They are nearly 1/4 inch in diameter. You should get along very well with 3/16-inch ball bearing balls.

By the way, the many groups who have seen my Franklin motor have often included someone who wants to know this: If a Franklin motor will run, why can't large electrostatic motors be made and used? Why do we have vast numbers of *electromagnetic* motors, from tiny to immense, but no practical *electrostatic* motors? The answer hinges on the fact that electrostatic forces are little forces. To increase them, you need electric fields of high intensity. But when you go too high with your voltage and high intensity, you get wasteful corona, or spark-over. To stop that, you have to separate the parts some more. But that reduces your intensities and your forces, and you have to go still higher in voltage. Soon, you find that you are just up against it: inherently, any workable electrostatic motor would have to be far larger, and far more costly, than an electromagnetic motor of the same horsepower output.

The Dirod as a Motor

In theory, a Dirod should run (backward) as a motor if another Dirod feeds it as a generator. Since I knew that the electrostatic driving forces are so weak compared to machine friction, I confess that I had never tried it. Then, on November 2, 1966, 1 gave a lecture-demonstration to a couple of hundred very sharp research men at the National Research Council of Canada, in Ottawa. Then I turned them loose with the apparatus. Later, to my delighted surprise, I found one group actually driving one Dirod as a motor, feeding it from another. True, it barely turns, and only one or two of my family will really motorize. But that demonstrates the principle.

The Interdigital Motor

This motor, shown one-third full size in Figure 20 and in Plate 7 on p. 75, is a glass bowl with strips of aluminum foil epoxy-glued to it. Six of the strips are positive, six negative, and they alternate. The lug at the right connects to a strip running far enough each way, around the inside of the top of the bowl, to connect to the strips that go down in but stop short of the center. The lug at the left connects to a strip that touches the others at the center. Silver paint connects the lugs to the strips.

With lugs connected to a generator, I throw a conducting ball in along the sloping side to start it, whereupon it chases around and around, often getting up enough speed to climb halfway up. More balls can be added, one at a time. One lucky day I had thirty-one balls whizzing around in that bowl!

When a ball rolls onto a positive strip, it becomes positively charged, is repelled by that strip, and is attracted by the next negative strip. When I dreamed up this new device, I did not foresee that it would become so much of a favorite.

Here again, we have a ball that is a charge-carrier. When it cannot take care of the generator output, the voltage rises to where there is spark-over be-

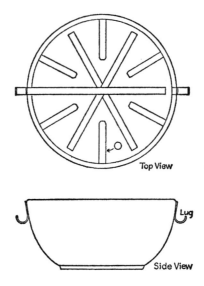

Figure 20. The Interdigital Motor.

tween strips somewhere, and the behavior becomes erratic. The best performance is to throw off the belt, turn the generator very slowly by hand, and feed the motor without overfeeding it.

Smaller aluminum balls work very well, compared to the special balls I like to use (those copper-coated polystyrene balls). Some smooth lead shot I picked up somewhere, a little over 1/16 inch in diameter, also do very well.

The Plateful of Balls

This is a very popular demonstration. The Melamine dinner plate, Figure 21 and at the right in Plate 7 on p. 75, has flat bottom and smoothly curved sides. It rests on an insulating slab. Two lengths of pull-chain form the electrodes. The balls are those "specials"—copper-coated polystyrene, but other balls will work. With the chains connected to a generator, the balls dash back and forth, carrying charges from one electrode to the other. A pasteboard band an inch or so high is put all around the plate, to keep balls from shooting off.

Now, going back to Chapter 7, where we had corona at both electrodes, with plus ions streaming one way and negative ions going the other way, and that discussion about why the two kinds did not at once get to meet and recombine—here you have a demonstration of that. A positive ball crossing left comes near a negative ball crossing right. They *may* meet and discharge to

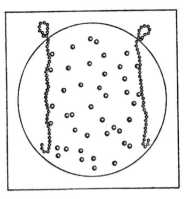

Figure 21. The plateful of balls.

each other; but, depending on their paths and speeds, they may merely swing past and go on. Not only that: other balls are in the way, some charged, some uncharged.

You can easily prove that such recombination meetings do occur. Operate the plateful of balls in a dark room, and watch. You will see little flashes here and there, where the oppositely charged balls meet, discharge, and neutralize each other.

One of my ambitions for some years has been to make the balls circulate around the plate, somewhat as a Franklin motor rotates about its axis. For this, I have tried removing the chains, and pointing two rods down at the balls in various ways. I have never succeeded. That's something you might try. Good luck!

The Ball Box

The box is shown one-third full size in Figure 22, and also near Dirod I in Plate 7 on p. 75. The bottom and ends are Plexiglas, epoxy-glued to the aluminum sides. With electrodes energized, aluminum balls bounce back and forth at high speed. Steel ball bearing balls, 3/16-inch or smaller, will work very well. Here again, we have balls carrying charges.

The Vertical Ball Dance; and Separation

If balls can be made to bang back and forth, why not try to use smaller balls and make them go up and down? We surely can. Figure 23 shows the box, one-third full size; and in Plate 4, Dirod Junior is ready to feed it and make the balls perform. The box has aluminum plates for top and bottom,

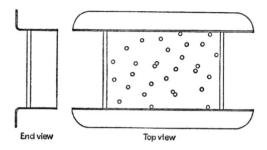

Figure 22. The ball box.

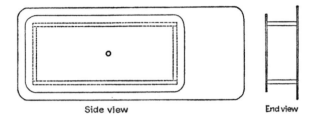

Figure 23. The vertical dance of balls.

with sides of Plexiglas. Several hundred 1/32-inch steel balls put on a splendid display, with the Coulomb force banging them up and down at high speed. A great many viewers have been fascinated by the ball dance. When you come to electrocoated sandpaper in the next chapter, you will recall this dance.

But this isn't all. Mixed in with the steel balls, I have a lesser number of sapphire ball bearing balls of the same size. Next, see that little hole in the upper plate (which is removable)? Balls come shooting up through it. They would be lost, of course, so we have a trap: a ring of some sort, with a glass top. And now we see a beautiful effect: it is the *steel* balls that come up! They, being conductive, acquire a charge and swing into action. The sapphire balls, being nonconductive, are passive. Yes, they do get kicked around by the other balls; and once in a while, by rare accident, one does get kicked upstairs. So here is one case of *electrostatic separation.*

One tour took me to Miami on a hot, humid day, and the location meant plenty of salt particles in the air. That was the day when the sapphire balls went into action too, and came up. Their surfaces had become somewhat conductive.

There's a lesson in that! And the workers in the electrostatic separation area have learned it long since: if a separation process won't work because one of the kinds of particles is too nearly nonconductive, they must just find some cheap chemical treatment that *will* make the surface somewhat conductive.

The Electrostatic Separation Industry

In Chapter 6, two different schemes for separating a mixture of two kinds of particles were mentioned. A third has just been covered. Maybe these examples have already set you to wondering if you could work up your own stunt for showing electrostatic separation. If not, perhaps what follows will do so. You may already realize that the workaday world must offer many cases where separations have to be made, and that electrostatics may furnish *an* answer, and sometimes the *only* answer. If so, you are right. These needs exist. And long before much of the theory of ionization by corona, and frictional or contact electricity were fairly well worked out, ingenious and stubborn men were hard at work, experimentally devising useful methods and machines for electrostatic separation of many mixtures. Invention and improvement continue, for this field of application is wide open.

There is dirt and such to be removed from food products, as in the cleaning of mustard seed. The clearing of weed seeds from good seed (such as timothy) needed for planting, is another instance. Other natural products, such as gums, resins, shellac, cocoa beans, require cleaning. Very large amounts of such materials are electrostatically cleaned.

The mineral industry has presented man with a huge and highly varied job to do. In one kind of case, ore-bearing rocks are crushed and ground fine. The valuable ore particles may then be separated out from the worthless rock particles by flotation, by magnetic separation, or electrostatically. Sometimes it takes a combination of two of these, or even all three, to get the job done. Or again, the mixture may *occur* in particle form in nature, as in Florida, Australia, and many other places. The problem is to handle immense tonnages of worthless sand cheaply, discard the sand, but retain some valuable ore—such as titanium ore.

In all cases, there must be one or more *differences* between the kinds of particles to be separated: difference in size, or density, or surface conductivity, or dielectric constant (to be taken up later), or contact potential property, or even other differences not mentioned. Some processes work without using corona and ionization; others require it. As you sense, we can only hint here at the wide range of possibilities, and at the different ways in which highly successful electrostatic separations are achieved. Although many patents have been taken out here and abroad in the past eighty years or so, and many papers have appeared, it is notable that there is but one comprehensive book on the subject—*Electrostatic Separation of Mixed Granular Solids*. It is by a man long with the U. S. Bureau of Mines, a man who was one of Cottrell's early associates—O. C. Ralston. Thumbing through it quickly indicates the complexity and diversity of electrostatic separation, and its great possibilities.

J. Hall Carpenter, founder and President of Carpco Research and Engineering, Inc., informs me that iron-ore separation at the Wabash plant in Labrador amounts to 2000 tons per hour; that the Trail Ridge plant in Florida handles 200,000 tons of material per year; and that his company's machines alone, installed in a great many countries, electrostatically treat 10,000,000 tons of minerals per year. By keeping these weak little electrostatic forces busy enough, tremendous quantities of mixtures are separated, and a job of prime importance to the world's economy gets done.

Demonstrations with Liquids

Although this book holds mainly to open-air electrostatics, I have already mentioned Herbert Pohl's work, much of which is with liquids. Some more striking demonstrations with liquids are presented by C. L. Stong in The Amateur Scientist section of *Scientific American* for January, 1967. Look this up, by all means. Here, Stong presents experiments worked up by Roger Hayward, illustrator for that department of the magazine.

A number of the experiments are done with carbon tetrachloride. The article has several warnings about the poisonous nature of this liquid, but in my opinion, the warnings are not strong enough. It comes to this: unless due precautions are *sure* to be taken, "carbon tet" *must not* be used indoors. Inhaled in sufficient amounts, the stuff is insidious and deadly—all the more so, in that people who work with it day by day are being poisoned without suspecting it. Then the crippling effects finally strike, and as far as I know, there is no cure for the damage done. Therefore, completely avoid inhaling the fumes; have the room *well* ventilated; if the stuff gets spilled, get out until the air is safe again to breathe. Also: "carbon tet" is far too commonly used by uninformed people. Be a good citizen, and never hesitate to pass this warning along.

One Type of Electrostatic Separator

A type of separator widely used is shown in Figure 24. It is not drawn to scale, and the particles are greatly exaggerated in size. A hopper above feeds the mixture of two kinds of particles, conductive and nonconductive, onto a rotating steel drum. The positive side of the power supply is connected to the roll, which also is grounded. At the upper right, you see the end of a wire, sus-

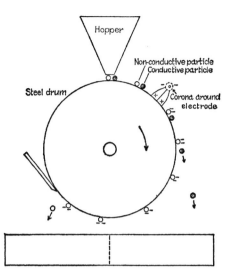

Figure 24. Schematic view of electrostatic separation of mixture.

pended parallel to the roll. It is the negative electrode. Enough voltage is used to make corona around the wire. The negative ions flow along the field lines to the roll. The conductive particles are the black ones.

Plenty of ions gather onto the conductive particles, but the particles conduct this charge to the roll. Such particles remain at roll potential and soon fall off—or even may be propelled off.

On the nonconductive particles, the ions largely gather on the *outer surfaces*, remain there (at least briefly), and cause these to be *attracted* to the positive roll. They are *pinned*, as the saying goes, and are scraped off at lower left.

Separation without Corona

We must recall here Herbert A. Pohl's research, already discussed in Chapter 6, in which advantage is taken of the tendency of particles to move to the more intense regions of the nonuniform field. Heretofore neglected, this relatively new development may offer rich possibilities as more workers get into this area to see what can be done.

10. More Service from Corona: Electrocoating

We have considered two of the great areas of applied electrostatics: precipitation, and separation of mixtures. Now comes the third: electrocoating, both dry and wet. The term itself can be subject to misunderstanding.

Ordinary electroplating by which, say, chromium is plated onto an auto bumper, could be called electrocoating, but is not. And again, there is a different process, old in some respects, but only now making progress industrially, by which, for example, a whole auto body is immersed in a tank of water. The water has resinous solids suspended in it. A current is passed between electrodes and the body (by the way, in the opposite direction to that used in metal plating). And, by what is called *electrophoresis* (plus two other effects) a primer coat of great uniformity is coated onto the body. This not only *could* be called electrocoating, but *is* so called, to some extent.

As taken up in this chapter, electrocoating is strictly confined to electrostatic processes.

Electrostatic Sandpaper

Coated abrasives, as they are called, include sandpaper, sander disks and belts, grit papers and grit cloths, and so on. The grit, which may be anywhere from extremely fine to very coarse, may be iron oxide, aluminum oxide, emery, garnet, carborundum, or other hard, sharp type of particle. In Chapter 16, where the making of the Junior Dirods is taken up, it will be necessary to describe the use of some of these products.

The old way—and the easy way—to make sandpaper, for instance, is to spread glue on the paper, sprinkle on the grit, and let it dry. Some of it is still made that way. However, the grit particles, being irregular, tend to lay down on their sides. With many abrasives, the product would be superior if they could be coaxed to stand up, more or less end-on.

The electrostatic way of making better coated abrasives is shown schematically in Figure 25. There are two metal plates as electrodes, with a suitable high voltage between them. Above, the paper, with the adhesive applied, moves to the right, rubbing on the upper plate. Below is a belt, moving to the right, carrying the grit, and rubbing on the lower plate. If the belt and paper and particles are sufficiently conductive, several things happen. The grit particles, on entering the electric field, stand up to align with the field. Next, they

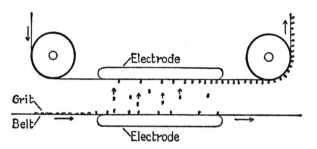

Figure 25. Making sandpaper electrostatically. Schematic, showing the grit behavior.

become charged and fly upward, just as the little balls do in my box of dancing balls. Next, they stick into the adhesive, more or less end-on.

And that isn't all. A great deal of "open-coated" abrasives are used, in which the surface is not "filled" with grit. Automatically, these particles, having like charges, repel each other on the way up, and tend to space themselves uniformly when they stick. On the other hand, if a *filled* product is desired, any particles coming too late to find an open space hammer the first arrivals all the more firmly into the adhesive.

Remember: my sketch is schematic, made just to show the principles involved. The actual machines are built on a slope, and they have little resemblance to what I show in the sketch. I have seen these coating machines at the Behr-Manning Corporation plant in Troy, New York, and believe me, they are giants—able to turn out a terrific number of square yards of sandpaper or other coated abrasives per year.

You would expect that some electrical engineer invented this process. Not so. As I have already indicated, much of the earlier invention and development of electrostatic processes came at the hands of ingenious and persevering men of varied backgrounds. And so it is here. In 1917, at the University of Michigan, I had, in one of my courses, a Civil Engineering student named Elmer Schacht. I knew him very well then, and still do. After college, Schacht went west and learned about electrostatic precipitation and separation. Then he joined the above company as a sales engineer. He invented the process for ·making electrostatically-coated abrasives. When he retired a few years ago, he was chairman of the board.

The output of coated abrasives in this country is probably around $125,000,000 per year. Much of it is made by Schacht's process. I understand that although the process is highly successful, and that much research has been done on it, questions still remain as to the intimate details of what goes on. As these questions are answered by more research, still better processes and still other products may be the outcome.

Carpets, Upholstery, and Velvet Walls

Very few people have any idea of how much their lives have been affected by electrocoating. If grit particles can be made to line up with an electric field, space themselves apart, travel across a space, and stick end-on, why can't this be done to little fibers? It can be. The pile of some carpet is made that way today. I have read of the recent opening of a Georgia factory that makes carpet, 15 feet wide, electrostatically. In the industry, the process is called *electrostatic flocking*.

Velvets are made this way, with the pile applied everywhere on it, if you like. Or, for pattern work, the adhesive is printed on in the desired artistic formation.

More than that: the process, by way of a Detroit firm I have visited—Velvetex Industrial Corporation—can be carried right into your home, to give you velvet walls, ceilings, or even floors! First, the surface is given a coat of aluminum paint, to make it conductive; then the adhesive is applied to that. Now comes the operator with his power pack and applicator. The applicator carries tiny fibers of Nylon or other material, in the color and texture you want. Holding the device a few inches from the surface, the operator shakes it, the fibers tumble out, they are charged by corona made in the applicator, and—well, I wish you could see it! There they go, travelling end-on, straight to the adhesive, and sticking into it end-on. How many? Why, as many as 300,000 fibers to the square inch! The process can also put the velvet lining on musical instrument cases, on upholstery, and the like.

Nobody knows how far and how fast these new developments will go. If auto floor carpet were to be made this way instead of by the tufting process, it would call for 12,000,000 to 15,000,000 pounds per year.

Insecticides

The pollution of our environment is, as you know, one of our most pressing problems. Pollution of the only atmosphere we have, by factories, refineries, auto exhausts, and the like; pollution of our rivers and lakes and water resources by sewage and by factory discharges; pollution of some foods, and of waterways, the killing of many birds, by poisonous insecticide sprays and powders—all of these must be brought under control.

When a powdered insecticide is applied by a farm machine using air jets, much of it falls to the ground between the leaves of plants being treated. If there is a wind, you may see much of it being carried away, to fall elsewhere. In either case, there is waste.

But when the powder is applied electrostatically, the process tends to coat the leaves' *entire* surfaces, and not just the tops; thus, the application is more effective and less wasteful. I have visited the research headquarters for this activity, at the University of North Carolina at Raleigh, talked with the men who are doing the research, and have seen the devices. A gang of eight electrostatic dusting heads are spaced along a wide beam, to powder several rows of the crop at once, as the machine moves along. Not every kind of crop can as yet be so treated, but further work gives promise of expanding the applications. I was told that already there are some five thousand machines in use, chiefly in the South, Southwest, and California. Thus, one important area of man-made pollution is being seriously tackled.

Xerox

Xerox: a magic name in the history of American business. The name stems from *xerography*, which means "dry writing." The Xerox machine initiated a terrific revolution in the office copying field. It uses an electrostatic process. You lay a typed page, or a page of a book, face down on the glass. You push a button; and soon, out comes a copy, on whatever kind of paper you care to use. No wet chemicals are involved.

In 1935, Chester Carlson deliberately set out to invent a new, different, and better copier. After exploring other processes, he turned to electrostatics and the use of a *photoconductive* material. Such a material will retain a charge in the dark; then, wherever it is exposed to light, the charge disappears. In three years, Carlson had his first small success. He then approached quite a few companies with it, but, to their present great regret, they all refused to be interested. Not until 1944 did he arrange with Battelle Memorial Institute to do the development work. I had the privilege, a few years later, to be lecturing there, and got to see some very impressive results. Later, a company, which became Xerox, took it over, and the revolution was on. In Carlson, we have a stubborn man who met defeat time and again, but did not give up in all those years between the first invention and the success that has made him very wealthy.

Basically, a special form of selenium on a roll is uniformly charged by corona-produced ions. A bright light illuminates the original, and a lens puts the image on the roll. Where the light strikes, the charge is removed. Elsewhere (as where these words are in the image if this page is copied) the charge remains. A special dust is applied, which sticks to the words. It is electrostatically moved from the roll to the paper, and fused to the paper by momentary high heat. This is a very inadequate description—far from indicating the immense amount of research and development it took to make the process a commercial success. I invite you to take a look inside of a Xerox machine. It is packed so full of parts that a mouse could hardly find room to make a nest.

Electrostatic Printing

Xerox has quite recently developed high-speed machines that turn out twenty-four hundred copies per hour. This is more than just making a few copies: it is *printing*.

A different process is used by the giant publishing firm, Time, Inc., which has to print more than twelve million address labels per week. This is done electrostatically at the rate of 135,000 labels per hour, on a Videograph. Simon V. Bol has given a brief description of it. Think of a TV tube modified so that instead of having a fluorescent screen for the image, it has a face plate with a wire matrix. The electron beam "writes" the configuration of the address onto the wires. The charged wires transfer the image onto a moving web of paper, putting a patterned charge on the paper. Where the charges are, a suitable dust sticks until it is processed to stick permanently.

Electrostatic Spray Painting

Great quantities of mass-produced articles must be painted—toys, furniture, refrigerators, auto parts and bodies, and all that. Many would be costly if painted by hand. Spraying them with a spray gun is much faster. But in many cases it is very wasteful: all the spray that goes past the article is wasted. This is where electrostatics steps in to save great quantities of paint. And note this: the process is only twenty-odd years old.

One process is shown schematically in Figure 26. At the left is the spray gun, from which comes a cloud of very tiny paint particles—ejected from a small orifice by air or by hydraulic pressure. At the right is the target, represented by the three large circles. Imagine you are high up, looking down, and that a long series of these articles are being carried along past the spray zone by a conveyer chain. This, and the target articles hung from it by wires, are grounded; so also is the positive side of the power supply. The spray gun is the negative electrode, with anywhere up to 120 kV supplied. The dash line indicates the scope of the corona formed around the gun. All this might be true, if the gun *had* this shape, and had no points. I needed to draw it this way to have room to show the nearby particles. Actually, there would be one or more points sticking out from the gun. Most of the corona would form on these points.

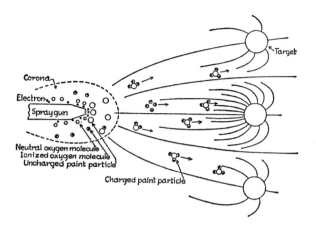

Figure 26. Schematic view of electrostatic paint spraying

Coming to the various details I have shown, let us sort them out. The black dots are electrons, coming in vast numbers out of the negative electrode surface (the gun). The little open circles are ordinary, or neutral, or nonionized oxygen molecules of the air. The little circles, each with a black dot, are ionized molecules—that is, ions. The somewhat larger open circles are paint particles. The plain circles are particles not yet charged. But out in the open space in the electric field, you see some that have picked up two or three or four ions, thus becoming negatively charged.

Remember that this is a schematic drawing. The electrons and molecules and paint particles have been enormously magnified. Instead of a few such, there are millions and millions of the first two; and no doubt, hundreds of thousands of the paint particles. I have shown only a few ions attached to the particles; instead, there could be a very large number.

So, what happens? The lazily moving cloud of particles drifts to the target for three reasons. The gun gives them some velocity; the charged particles are urged along the field by the Coulomb force; and there would be an electric wind (with small effect, no doubt). Now look at the shape of the electric field near the articles. Those tiny charged particles tend to follow those lines, even if it means coming around a curve behind and landing on the rear of the article. Yes, they actually do! This is called "wraparound." By it, much paint that would go past and be lost, is deposited on the article. If you wonder how the ions manage to gather so well onto the paint particles, go back to Figure 9 and see again how objects placed in a field gather the field lines into themselves. The ions thus are guided to the particles.

And now, *space charge* again: all those charged particles in the field space, along with unattached ions, certainly do charge the space. This definitely has its effect on the whole set of phenomena, both on the complete understanding of what goes on and why, and on the design and operation of the spraying setup.

If the corona at a negative electrode, with a current of only 1 microampere (a millionth of an ampere) produces 6,230,000,000,000 electrons per second, each ready and willing to make an ion think of the flood produced in paint spraying, which may go to 10 or 20 milliamperes. At 10 milliamperes, electron production is 62,300,000,000,000,000 electrons per second.

The above process, which depends on corona, is used by the spray guns made by the DeVilbiss Company of Toledo, Ohio, and by a hand gun made by the Ransburg Electro-Coating Corp. of Indianapolis, Indiana.

Also in wide use are electrostatic sprayers made by Ransburg that do not depend on corona-produced ions to charge the particles. In one form, paint is fed onto a thin rotating disk, through a hole at the center. The shaft is vertical, the disk horizontal. If operated without voltage applied, the paint would be slung off the edge in irregular strings and blobs. But when we apply, say, 120 kV, with the disk negative and the nearby articles grounded, the paint forms very fine, regularly spaced tiny streamers, from the ends of which very fine droplets come off. Now, solvents used make the paint *conductive*. So, first, the electric field acts to make those streamers; and second, as in the Kelvin generator, the droplets leave, negatively charged—and are impelled along the field to the painted articles. The articles are marched around the disk in a nearly complete circle by a conveyer chain. In another form, the disk is replaced by a bell, which faces the work that hangs from the chain. The paint saved depends on a number of factors. It ranges from a considerable saving, up to where there is practically no waste. Dr. Emery P. Miller, Ransburg's Vice President, Research and Development, informs me that in one example, spraying twenty-four hundred washing machines by hand spray gun, not electrostatic, used $900 worth of paint. Changing to electrostatic painting made a saving of $540 per day. One rough estimate is that in the U. S., electrostatic processes may save paint worth $50,000,000 per year.

11. Fun with Capacitors

In the years between von Guericke and Benjamin Franklin, the Leyden jar was invented—invented independently, by the way, by E. G. von Kleist (1745) of Pomerania, and P. van Musschenbroek (1746) of Leyden. In all of the literature of the electrical sciences published before 1850, it may be that the Leyden jar was mentioned more often than any other piece of equipment.

It apparently was the first simple and reliable *condenser*. In its final form, it was—and is—a glass jar, coated part way up, inside and out, with tinfoil. There is an insulating lid, with a rod down through it. There is a ball on the top of the rod; and a chain hangs from the rod to make contact with the inner coating. By charging the jar, electrical energy could be stored ("condensed") and later released. Experiments in the early days were by no means confined to sober research. Bern Dibner, in *Ten Founding Fathers of the Electrical Science*, tells how Louis XIV got to see 180 soldiers of the guard jump simultaneously when, holding hands, they took the discharge from Leyden jars. There was another time when "700 monks of a convent in Paris" likewise jumped in strict unison. (Were you wide awake when you just now read about "monks of a convent"? Make you wonder? I leave it to you to do as I did: look up the word "convent" as used in those days.)

Until recently, the name "condenser" was still in wide use. Today, the swing has already largely taken place by which these are now called *capacitors*. This is true in this country. I believe that Britain still prefers the old name.

What Is a Capacitor?

In the figures in this book, you have already looked at a number of capacitors, perhaps without knowing it. Any two electrodes, surrounded by empty space, or air, or good insulation of any kind, constitute a capacitor. For they can be given opposite charges, the charges set up an electric field, and there will be energy stored in the field. To be personal about it, if you and your best friend stand on an insulating floor, you two are a capacitor. Let each of you touch a terminal of a Dirod or Wimshurst or Van de Graaff: you are oppositely charged, and you set up a field between you. Take your hands from the terminals, and you are still charged. Now start to shake hands. Before you touch, a spark jumps between you. Some of the energy of the field appears as heat, and some is radiated from the spark.

Let us take the pairs of electrodes in Figure 3, Figure 6a, and Figure 6b, in this order, and treat them as capacitors. Think of the rods of the first two pairs as being many feet long (into the paper), and of the plates in the last one as also being long (into the paper). Now compare them in terms of 1-foot length of each case. In the first, we have small rods, quite far apart. A given voltage applied to them gives them a weak charge and sets up a weak field. The second pair's rods are larger and closer together; the same voltage puts on larger charges and makes a stronger field. In the third case, the plates are wide and close together: the same voltage gives a much larger charge and much stronger field.

The ability to store charge, for a given voltage, is called *capacitance*. The capacitance of those three cases ranges from small, to considerably large, to very much larger.

When a capacitor discharges by way of a spark the current in the spark is typically *oscillatory*. It surges back and forth quite a few times at high frequency before it dies out. This happens too fast to see.

Amount of Capacitance: the Farad

The *unit* of capacitance, the *farad*, designated by the letter F, is obviously named for Michael Faraday. It so happens that this is a *very* large unit. One whole farad is such a large capacitance, compared with those in use, that you will probably never get to see a 1 farad capacitor. A thousandth of a farad, or *millifarad* (symbol, mF) is a smaller unit; smaller yet is the *microfarad*, a millionth of a farad (symbol, μF). But we need a still tinier unit, the millionth of a millionth of a farad, called a *picofarad* (symbol, pF). Incidentally, if all this makes a picofarad sound so small that you tend to lose interest in it, don't. Wait until you take a poke from, say, a 150 pF capacitor charged to 40 kV. It would not harm you, but it certainly will keep you interested.

At the end of this chapter, we will see how voltage, charge, capacitance, and energy are related; but these affairs will mean much more to you if we first consider some actual capacitors. We will do that after taking a look at *dielectrics*.

Dielectrics

If the parallel-plate capacitor in Figure 6b is measured in a *vacuum*, and then in *air*, we would find the capacitance to be practically unchanged. The air molecules of nitrogen, oxygen, a little carbon dioxide, a very small content of argon and such, have not affected the situation. Nevertheless, the air is there, and air is an insulator (except when breakdown occurs), and it is properly called a *dielectric*. It has what we call a *dielectric constant*, by which we compare it to a vacuum, or pure space. Obviously, since air and vacuum gave the same result, the dielectric constant of air is 1, or unity.

Now get a slab of almost any good insulating material and cut it to fill the space between these two plates. Use Plexiglas, polyethylene, beeswax, paraffin, solid polystyrene, or glass—or even try insulating oils or other non conducting liquids—and again measure the capacitance. Now, all these are insulators, preventing current flow between electrodes; and in that sense, they are all dielectrics. But next comes the good news: depending on what "filler" you use, you might find that the capacitance had been anywhere from doubled, on up to being made eight times as much. That is, many insulators have dielectric constants of from 2 to 8. Some are higher; *pure* water has the amazing value of 81 for its dielectric constant.

Low-voltage capacitors need very little insulation; they can be made more compact by using an insulant with a high dielectric constant. *High-voltage* capacitors can be made (and some are) with air as the dielectric; but wide spacing, needed to prevent breakdown of air, makes them large. Others use a good dielectric. It must be thick enough to avoid being punctured, and extend far enough beyond electrode edges to prevent spark-over. With a dielectric, the electrodes or plates are brought together as close as can be, thus getting maximum capacitance per unit of area.

Why do insulators insulate? Why are they nonconductors? Going back to conductors (Chapter 4) you recall that their atoms release electrons, so that there is an electron cloud, ready to drift when urged. When drifting through the conductor, they constitute a current. Or when electrons appear in excess at a surface, and there is a deficiency elsewhere, we have oppositely charged surfaces.

Insulators are different. Their atoms or molecules are happy the way they are, and tend to retain their electrons. But they also have another property: their atoms or molecules can be *polarized*. Even though neutral as a whole, the negative electrons in orbit can be lopsided with respect to the positive protons in the nucleus, giving the effect of a negative charge at one side and a positive charge at the other. When an electric field is applied, these little polarized units tend to line up with the field. The net effect is to enable the electrodes of the capacitor, when the space is filled with one of these dielectrics, to have much larger charges placed on them by a given voltage.

Air as an insulator: is it perfect? No. Suppose we hang a smooth metal ball in air on a quartz fiber and place a charge on the ball. We can ignore the leakage by conduction along the fiber. Then why should the ball ever lose its charge? We have already seen that radioactivity and cosmic rays maintain an ion supply in the air. If the ball is, say, positively charged, then negative ions will drift to it and gradually neutralize it. There seems to be a belief, very widely held, that moist air leaks more freely than dry air. There seems to be no evidence to support this belief. Undoubtedly, it stems from the fact that *surface* leakage is often much affected by high humidity.

Speaking of dielectrics, a dielectric was used in a great favorite of the early experimenters: the electrophorus. Very simple, very interesting, and you may want to make one. You will find it written up in a good encyclopedia.

A Kitchen-made Capacitor

Can you walk into your kitchen and make a good capacitor? In this Age of Plastics, you certainly can. All you need is a bowl of *any* kind, a smaller plastic receptacle, and some water. My kitchen capacitor, shown one-third full size in Figure 27, uses a polyethylene Freezette for the inner receptacle.

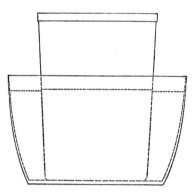

Figure 27. Kitchen-made capacitor. Two containers and some water.

Freezettes are made for keeping food in a refrigerator. Polyethylene, a tough, milky, stiff, but somewhat flexible plastic, is widely used in kitchen articles and as an insulator in electrical apparatus of some types.

With the water height at about 3 inches, this capacitor has about 53 square inches of useful surface. It then has a capacitance of about 100 pF. In a Leyden jar, the inner and outer tinfoil coats are the electrodes, and the glass is the dielectric. Here, the inner and outer *waters* are conductive, and *they* are the electrodes. The polyethylene is the dielectric. This capacitor went to 60 kV before sparking over.

Before going on, let us come back to this business about the water being conductive. Any tap water is impure. The salts dissolved in it form positive and negative ions, and these are ready to move anywhere, when urged. They are movable charges, just as the electrons in the electron cloud in a metal are movable charges.

The Horizontal Capacitor

If we want a rugged capacitor for experimenting, the simplest form for it is a three-piece job: a Plexiglas sheet with an aluminum plate on each side of it. Very simple indeed. *But not useful.* If we lay it flat, we cannot easily connect to the lower plate. If we stand it up on edge, there are two problems: how to hold it, and how to connect *easily* to the two plates.

I designed the horizontal capacitor shown one-sixth full size in Figure 28, to be laid flat and to permit *both* connections to be made *on top.* You may see

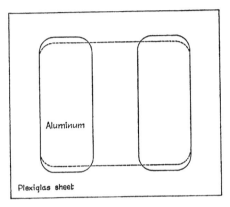

Figure 28. The horizontal capacitor.

it in the center of Plate 7 on p. 75, connected to the Radial Dirod. The Plexiglas sheet, 12 by 14 inches, is 1/8-inch thick. The plates are 1/16-inch aluminum, with corners rounded to a one-inch radius, and edges rounded and smoothed. The lower plate, 8 by 10 inches, can be temporarily taped to the Plexiglas if you are in a hurry (what experimenter isn't!); better yet, use epoxy at the edges. The two top plates are 3-1/2 by nearly 9 inches. With the top plates placed as shown, this capacitor has about 50 pF.

In use as shown in Plate 7 on p. 75, there are really *two capacitances in series*, each of about 100 pF. Suppose the Dirod delivers 40 kV. Then we have 20 kV from one top plate to the bottom plate, and the other 20 kV from the bottom plate to the other top plate.

Notice that the Plexiglas extends 1-1/2 inches or so beyond any plate edge. With less extension, there might be spark-over at roughly 40 kV. With the top plates located as shown, *incipient breakdown* at around 40 or 50 kV can usually be heard, and in the dark can be seen: a beautiful display of little flashing streamers between the inner edges, and perhaps elsewhere, along the

Plexiglas surface. It may lead to a spark between the inner edges. If not, sliding the plates closer together will bring the spark.

With the top plates loose, we have an *adjustable* capacitor. Placed as shown, it is at maximum capacitance, and if you bring your knuckles close to both plates, 40 kV will give you a stimulating but perfectly safe shock. As you slide the plates apart and reduce the effective areas, the spark intensity and the shock are reduced. Due to incipient breakdown, or spark-over, this capacitor will not consent to go above 40 to 60 kV.

The *clapper* (Chapter 9) can give a very convincing demonstration of the fact that the greater the capacitance, the greater is the energy stored at a given voltage. First, the clapper alone is connected to the generator and operated. Suddenly stopping the generator does not immediately stop the clapper. It continues for two or three or more weakening swings, as it drains the energy stored in the low capacitance of the generator and connected parts. Next, connect the horizontal capacitor to the generator and the clapper. Operate. Again stop the generator suddenly. The clapper now goes for a much longer time, running down the much larger energy of the larger capacitance of the system.

Sphere Gap Discharge of Capacitor

With a generator feeding the horizontal capacitor, and also connected to the sphere gap, repetitive spark occurs at the gap. This makes a fine demonstration. The rate of sparking depends, for a fixed gap, on generator speed. Double the speed, and you double the spark rate.

As already indicated, these sparks are not just a sudden one-way flow of large current: they are oscillatory. This rapidly reversing current sends out *electromagnetic waves*, which travel at the speed of light. Heinrich Hertz, in Karlsruhe in 1888, first demonstrated this phenomenon, and this was a major breakthrough. It verified Maxwell's prediction of these waves as an outcome of his theoretical analyses. At some distance from the discharging setup, Hertz placed a loop, closed except for a small spark gap, and he had sparks jumping that gap. For more on electromagnetic waves, see John R. Pierce's book *Electrons and Waves*.

Separately Excited Dirod

The Dirods described herein are all *self-excited*. Therefore, when the spark jumps the gap as above, and largely discharges the system, the *induction of charge* goes down with the system voltage—and then builds up again as the system voltage builds up.

Why not *maintain* the induction of charge, by maintaining the voltage on the inductors? I have enjoyed doing this, by operating a Dirod that is *separately excited*. First, I disconnect the inductors from the collector plates, and rig up the inductors with insulative supports. Then I connect them to *another* Dirod, which is now the *exciting generator*. Then, in the repetitive sparking demonstration, the sparking rate is roughly doubled.

Dirods in Parallel

In my laboratory sometimes, when I crave lots of action, I have fun running two Dirods in parallel. Either terminal of one is connected to a terminal of the other; then the remaining terminals are connected. Connection is also made to the capacitor and to the sphere gap.

With both running, there is a high spark rate. When I shut one generator down, the rate of sparking drops. As it is speeded up again, the rate rises. Thus I play back and forth with them, enjoying the pleasant speed-sounds of the generators, and the cracking of the sparks. Once, I did put four Dirods in parallel—and you should have heard it then! (All this is by way of persuading you that after you build one Dirod, you really won't be happy until you give it a playmate.)

By now, you may have a question. If we have, say, two Dirod Juniors, each willing and anxious to make 60 kV, why not hook them in *series* and get 120 kV? We would try this by placing them side by side, connecting the adjacent terminals, and using the outside terminals as our 120 kV source. Yes, I had this happy idea. And no, it doesn't work. If the two connected terminals are, say, positive, the outside terminals insist on being negative, and the two voltages oppose each other.

The Roller

Lay a length of 1/4-inch aluminum rod, rounded at the ends, between the plates of the horizontal capacitor. If the charged capacitor is quite level, the rod rolls back and forth between plate edges as a charge carrier; and it may get up enough speed to roll up over an edge and roll off.

The Rockers

My demonstration audiences are invariably delighted with the rockers. Part of the stimulation for dreaming them up came from the flapping capacitor, discussed later on. Figure 29 shows one rocker one-third full size; and the horizontal capacitor supporting both rockers, 1/12 full size. You also see this combination in Plate 7 on p. 75. A rocker is about 1 inch wide, made of 1/16-

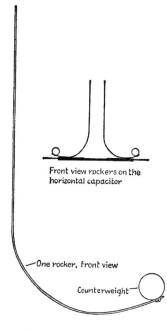

Front view rockers on the horizontal capacitor

One rocker, front view

Counterweight

Figure 29. The rockers.

inch aluminum, and with a suit able counterweight attached at the outer end. Using your own materials, you will need to experiment with rocker curvature, counterweight, and spacing between rockers, to get the best effect.

Placed on the capacitor plates, and with a generator feeding the combination, the rockers swing together, with a bright spark zipping between them before they touch. The opposite charges that attracted them together are now largely gone, and they are free to rock widely apart. But while they are separating and then starting to return, the capacitor is again being charged, they are pulled together again, and again produce that spark. Action, and sparks! Everyone loves this outfit. Everyone laughs out loud at it—and I don't know why. But we do!

As I have said before, when you try new things with electrostatics, be prepared for surprises. Sometimes a bright idea ends in total disappointment. Then again, the completely unexpected can happen—and there is a new demonstration for you. These rockers certainly handed me a surprise. One day, the rockers inched their way along toward each other, making shorter and shorter swings, until they came right together—and stuck together. Now, that seemed impossible. Why? When I shorted the generator by putting fingers on the top plates of the capacitor, the voltage went to zero, and the rockers swung apart to their natural, gravity-imposed positions, as they should. But when I removed the short, they again swung together and clung to each other. What went on here? When touching like that, they themselves should short the system, and the voltage should be zero, and there should be no attraction...! Do you see the dither I was in?

This had to be looked into. And close observation revealed two things. First, they were very slightly but very rapidly *vibrating*—so little and so fast that it was hard to see. Second, down in between somewhere, one could see a tiny discharge. So, *electrically* they were not in contact. At the very few places where they did *mechanically* touch, the aluminum oxide, which is an insulator, always present on aluminum, apparently insulated them from each other effectively, at the very low voltage prevailing. The rockers had turned themselves into a vibrating, discharging system. Of course, I promptly added this to my bag of tricks. Many a highly trained observer has been completely mystified by it.

Capacitor Relationships

Now that you have some idea of 100 pF in a kitchen capacitor and 50 pF in the horizontal capacitor, it is time to develop capacitor relationships.

The unit amount of charge is called a *coulomb*. Suppose you rig a flashlight cell as part of an electrical *circuit*, so that through a wire, it makes 1 *ampere* flow steadily in the wire, from one cell terminal to the other. When that *current* of 1 ampere flows for 1 *second*, a certain amount of *charge* has passed any point on the wire. It is called an *ampere-second*, and the other name for it is a *coulomb*.

Of course, what really happens is a very slow drift of the electron cloud within the conducting wire; and therefore, a certain number of electrons have passed that point. Since the charge on one electron is only 1.602×10^{-19} coulombs, it means that an ampere-second, or coulomb, represents the passage of 6.23×10^{18} electrons; or, writing it out, 6,230,000,000,000,000,000 electrons.

The symbol for the *amount* of capacitance is C. The symbol for the *amount* of voltage is V. The symbol for the *amount* of charge is Q. Here is an equation:

$$Q = CV$$

What does it say? It says that the coulombs of charge on each capacitor plate equals the amount of capacitance in farads times the voltage in volts.

Very well. Here we have a most useful equation. But it doesn't tell us how much *energy* is stored.

To work up to that, we must have a *unit of energy*. Consider a lamp of the ordinary type—a light bulb. Suppose that 100 volts applied to it causes 1 ampere to flow through it. Electrical energy is being supplied to keep that tungsten filament hot enough to make it give out light. The energy is being supplied at a constant rate, and *rate of energy* means *power*. Now, *electrical power* is expressed in watts, and this is a 100 watt lamp. Why? The product of 100 volts times 1 ampere gives 100 watts. In one second, the 100 watts would amount to 100 *watt-seconds*. And this is a certain amount of *energy*. Our unit of energy, a watt-second, has been named the *joule*. In one second, that lamp took 100 joules. How much energy will a capacitor store up—that is, how many joules? Another equation gives it:

$$\text{Joules} = \frac{1}{2}CV^2$$

Let us at once turn to SAFETY. It is known that a discharge into the human body exceeding *10 joules can be hazardous to life*, and that only 1/4 joule gives a heavy shock. Below is a table showing several values of capacitance. Under each, you find the kV that would store up 10 joules in the capacitor:

Table 2. Voltage to store 10 joules in a capacitor

microfarads (µF)	0.002	0.20	20	80	320	2000
kilovolts (kV)	100	10	1	0.5	0.25	0.1

By all means, get some practice: use the equation above and verify the values in the table.

Since human lives depend on being right about these matters, I am under obligation to issue a warning here. The above information came to me in a safety bulletin issued by a famous university. I found this: the capacitance values were stated to be "F" (in farads), and not, as you see above, in "µF." The omission of that little symbol "µ" in a safety bulletin is a deadly mistake. Furthermore, this table had been distributed by the U. S. Atomic Energy Commission, which had obtained it from the United Kingdom Atomic Energy authority. *If* the same mistake was made in those distributions, some tragically wrong safety information has been spread far and wide.[*]

---

[*]One more thing—about symbols. Some new symbols are now in process of being adopted internationally. Some I use herein may be different from what you may see in future papers and books.

More about Certain Capacitors

If I charge my 50 pF horizontal capacitor to 40 kV, how safe am I if I take its discharge? Let's find out. Stating these amounts in formal fashion, $C = 50 \times 10^{-12}$ farads, and $V = 40,000$ volts.

$$\text{Joules (energy stored)} = \frac{1}{2}(50 \times 10^{-12}) \times (40,000)^2 = 0.04$$

Since 1/4 joule delivers a heavy shock, but is far below the 10 joules that might kill, it follows that 0.04 joule is much below the danger limit.

Next, does a Dirod, or other generator, have a capacitance? Surely. The collector plates of a Dirod, charged oppositely, amount to a capacitance; and other parts add some more. However, the plates are several inches apart; and, as you would by now suspect, the capacitance is quite low compared to the horizontal capacitor. Dirod I and Dirod II have about 5 pF. Dirod Junior, with plates closer together, may go to 6 or 8 pF.

One of the "classic" cases is the *isolated sphere*. Its capacitance is given by the equation $pF = 1.412d$, where d is the diameter in inches. You at once say: If the sphere is one electrode, or one side of a capacitor, where is the other side? In pure theory, we say, at infinity; at an infinite distance. That may not satisfy. So, think of the sphere as hanging anywhere in a room, but several diameters from any wall. The room walls will then be the other side, where the electric field lines will terminate. Then the sphere's capacitance will almost exactly be given by the equation.

We will now put the sphere equation to work in three cases.

First, if you have a Van de Graaff, measure the sphere and compute its capacitance. Go on to find the energy stored in joules if, say, it goes to 250,000 volts. You will find that the discharge to you will give a pretty good jolt, but that it is safe.

Second, find your own capacitance. Recently, a girl in high school (no doubt, with a project in mind) wrote to the University of Michigan to ask, What is the capacitance of the human body? And the letter sifted down to me for answering. There isn't any exact answer. To get a *rough* answer, estimate your body's volume somehow, but don't be fussy about it. Then find the sphere that has the same volume. Then find the capacitance of an isolated sphere of that size. This might come within 50 percent of a true answer. But look: suppose you go to the kitchen and stand between the refrigerator and the range. You are far from being isolated, and your capacitance would have a large increase.

Third, the earth is a sphere, nearly, and a mighty big one. May it be big enough to have a whole farad? Take its diameter as 8000 miles, and see. You will come out with only a part of a thousandth of a farad!—in fact, 0.72 mF.

Before we drop the joule, we should say how to pronounce the word. Webster says there is little doubt that Joule himself pronounced his name to rhyme with "cool." It also says that it is common in the English-speaking world to make joule rhyme with "jowl." Personally, I have seldom heard the second pronunciation—and I don't happen to like it. And again, how about *pico-*, in picofarad? Pronounce it "peek-o," and you will agree with the way I believe it is most often said in this country. Pronounce it to let the first syllable rhyme with "Pike," as in "Pike's Peak," and you will agree with what Webster says. This is all I know about it.

Unexpected Shocks from Capacitors

Follow me closely. I connect my sphere gap and my 50 pF capacitor to a Dirod. I open the gap wide, charge the capacitor, and then stop the generator. Slowly closing the gap, I observe that it sparks at 45 kV. Then I know that the capacitor had been charged to that voltage. All right. The spark has discharged the system, hasn't it? Maybe. We find out. I continue to close the gap, slowly. It sparks *again*, at perhaps 15 kV. There was a very considerable remaining charge! That oscillatory first spark died out before all charge was removed. So: if I want to discharge completely, I had better close the gap *completely*, and short-circuit (that is, "short") the system.

But there is more to be said, and the safety bulletin mentioned earlier says it. Many high-grade capacitors in use today will *recover* a fair part of the original energy storage *if left on open circuit after complete discharge*. They may do this in minutes, or in months, depending on the dielectric used. There may be as much as 10 percent recovery of the original voltage.

In your experimenting, you may decide to let off a really wicked spark by getting some large war-surplus capacitors. I don't recommend it. But if you do it take every precaution *while* experimenting. Run no risks. When done, be sure to *short* the capacitors, and put them away. *Don't* remove that short. If you do, this residual effect might give someone a terrific jolt, even months later.

12. Connectors; The Vertical Capacitor; Figuring Capacitance

In low-voltage work—as in wiring a few dry cells to a small lamp bulb—connections can be made with almost any kind of wire. Not so in high-voltage experimentation. Sometimes, wire will serve. It depends on such factors as how high the voltage is, how near it is to some part with opposite polarity, whether it is effectively insulated, and so on. Wire in the wrong place can spill a lot of corona.

Even when wire would serve, it can often be very awkward to take the time to use it, when it is necessary to change connections rapidly, or when one idea after another is being tried out. This is when suitable, *loose connectors* are a necessity. This chapter will describe connectors I have found to be very useful.

It will also describe a vertical capacitor, and some things that can be done with it. And it will tell you how to figure the capacitance of plate capacitors—with air between, or with a dielectric.

Connectors

Two pairs of connectors are shown, 1/16 full size, in Figure 30. They are

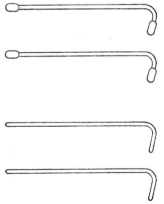

Figure 30. Connector rods.

made of 1/4-inch aluminum rods, bent at one end for hooking over a generator's terminals, or onto the sphere gap rods, and so on. The *upper pair* has knobs, to minimize corona leakage. These are of 1/2-inch aluminum or brass rod stock, turned and polished, and drilled for insertion of the connector rods. Epoxy glue is used to make sure the knobs stay on. These are for the higher voltages attained by three of the Dirod family—up around 80 kV or so. The lower pair, without knobs, have the ends rounded and polished. They are quite suitable for the lower voltages of the Junior Dirods, at around 60 kV or below.

Most of the demonstrations described in this book operate at 50 kV or below. Connectors for this range can also be made of 1/8-inch brass rods, with end knobs made of 1/4-inch rod stock. You can see all of the above, except the lower pair of Figure 30, in several of the Plates.

I also make much use of a straight pair, 16 inches long, shown in Plate 7 on p. 75, at center front.

Looking ahead to the vertical capacitor, Figure 31, it has four straight connectors that I have not shown in the drawing. They are knobbed at one end only. One pair is 6 inches long; the other pair, 9 inches.

The Vertical Capacitor

My *vertical capacitor*, with its auxiliary features, is the end product of much intense thought. It has turned out to be a device of considerable flexibility, and has given fine service. Shown about 1/7 full size in Figure 31 you also see it in service in Plate 3 on p. 42. The 1/8-inch Plexiglas sheet is 10 by

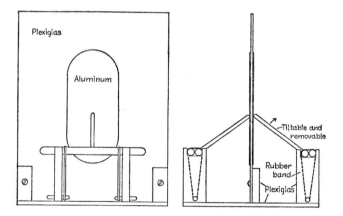

Figure 31. Combined vertical capacitor and connector assembly.

14 inches. Each aluminum plate is 3-1/2 inches wide by 7-1/4 inches long. The capacitance is about 160 pF. The main insulative structural parts are Plexiglas.

The pairs of 1/2-inch aluminum rods resting in recesses are snugged down by rubber bands. Of these, the *inner* bars are drilled part way in, to receive, *loosely*, 1/4-inch leaning connectors that reach up and touch the plates. The plates can be epoxy-glued to the Plexiglas sheet. To reduce plate edge corona, I applied the whitish stuff you see, which is RTV 102 Silicone Rubber—or corona dope can be used.

In use, the straight connectors described in the previous section are slipped under the rubber bands, as you see in Plate 3 on p. 42. The bands give that ready flexibility needed to accommodate the connectors to the levels of generator terminals, or the sphere gap, and so on.

This capacitor sparks over at around 70 kV. The spark reaches from one plate, along the surface, over the edge, and to the other plate.

And now, for a safety demonstration. Connect up as shown in Plate 3 on p. 42. Open the sphere gap *wide*. *Disconnect* the capacitor, by swinging the leaning connectors out from the plates by about an inch. Operate the generator for a while, and then stop it. Using a piece of insulation, push the leaning connectors back into plate contact. Slowly close the gap. Usually, for me, a 50 kV spark jumps the gap! *Without metallic contact*, the capacitor was being charged, unseen, by ionic flow from the leaning connectors, to the plates. Never forget this, if, someday, you are working around large, high-voltage capacitors.

Other Experiments

Disconnect the capacitor by removing the leaning rods. We now have a versatile connecting and electrode-extending device.

Place your hands in front of the knobbed outer ends, and feel the electric wind.

Use this setup to show the effects on the candle flame, as covered in Chapter 8.

Also use a knob for depositing that water drop described in Chapter 9, the one that is drawn out by the intense field, producing a fine spray.

Audiences invariably enjoy seeing a large Franklin-type motor. Such a motor is easily made. Cut a round disk from a manila filing folder, big enough so that when held near to the pair of front knobs, the knobs will be half an inch or so inside the disk. My disk is to the left of the horizontal capacitor in Plate 7 on p. 75. Provide a handle for it, a broken needle for a shaft, and a piece of Plexiglas for a hub. I recommend coating it on both sides with Krylon, to keep it from being too conductive in damp weather. Bring it up close to the electrodes, start it (either way) if you need to, and watch it spin. In the dark, you can see brush discharges between the knobs and the disk. This demonstrates that discharges are placing charges directly on the disk, where they are at least briefly retained. The action is the same as for the regular Franklin motor.

Figuring Capacitance

It is easy to compute the capacitance of two parallel plates with air between. Let A stand for the area of one side of one plate, and S be the distance between them, in centimeters. Then

$$C = 0.08854\frac{A}{S}, \text{ in pF}$$

If measured in inches, it is

$$C = 0.2249\frac{A}{S}, \text{ in pF}$$

Caution: strictly, these equations hold only for the area covering the uniform part of the field between the plates, somewhat in from the edges. However, if the smallest dimension of a plate is roughly ten or more times S, the equations will be pretty accurate.

Let's rig up a capacitor right now, and practice on it. And why not make a big one, while we are at it? Let the plates be square, and 10,000 feet each way. Place them 1000 feet apart. Using the second equation (and I leave you to verify it) we get $C = 0.27 \times 10^6$, in pF.

Now we apply a potential difference—and why be skimpy about that? Make it 100 million volts, or 10^8 volts, for fun. There ought to be quite a bit of energy stored. There is. And I leave it to you to check me out on this. Energy stored = 1,350,000,000 joules.

A bit more practice now: place those plates three times as far apart, or 3000 feet. That makes C one-third as much. In turn, it makes the stored energy one-third as much, or 450,000,000 joules. (These last two answers are pretty good, but approximate; for the spacing is large compared to plate size, and there would be a good deal of fringing of the field.)

Are you beginning to suspect something? Yes, that's right. We are getting ready to consider lightning, in the next chapter.

Capacitors with Dielectrics

When the space between the plates is filled by a liquid or a solid dielectric, the capacitance is typically increased. What you do is to look up the dielectric constant of the material, and multiply the capacitance figured as if you had *air*, by that constant. Here are some dielectrics, taken largely from

Attwood's book *Electric and Magnetic Fields*. For example, if a plate capacitor with air has 100 pF, filling the space with beeswax will raise it to 200 pF.

Table 3. Dielectric materials

	Dielectric constant	*Material*
(Gas)	1	Air
(Liquids)	25	Ethyl alcohol
	2.5	Transformer oil
(Solids)	4.5	Bakelite
	2	Beeswax
	2.5	Ebonite
	4.5–7	Glass, various kinds
	6	Mica
	4.1	Micarta
	2	Paper, dry
	2.3	Paraffin
	4	Plexiglas
	2.5	Rubber, pure
	4	Wood

Combinations of Capacitors

If we place capacitors in *parallel* and have them numbered C_1, C_2, C_3, and so on, the capacitance of the combination is

$$C = C_1 + C_2 + C_3, \text{ and so on.}$$

For example, you have eight capacitors, each of 20 pF. Your *capacitor bank* has 160 pF, connected in parallel.

Placed in series, it is more complicated; for three in series,

$$C = \cfrac{1}{\cfrac{1}{C_1} + \cfrac{1}{C_2} + \cfrac{1}{C_3}}$$

For example, those eight above, in series, would come out to give 2.5 pF. Or again, 4 pF, 4 pF, and 2 pF, in series would come to $C = 1$ pF.

A way to make yourself a powerful capacitor is to stack up the parts. You get three 1/16-inch Plexiglas sheets, say, 14 inches square. Then you make up three aluminum plates, 1/16-inch thick. Round the corners and round and smooth the edges. Now you stack them, with a Plexiglas first, on the table. Next, aluminum, then Plexiglas, and so on. You let the top and bottom plates stick out to the left, and connect them together. The middle sticks out to the right. Other than the parts that are sticking out, the plates everywhere have their edges in from the Plexiglas edges, by 1-1/2 to 2 inches, to keep spark-over from happening up to about 40 kV. You will have approximately duplicated my own capacitor of this kind. And believe me: I am very careful never to take a discharge from it. No, the energy stored is not really dangerous. But, although I am used to taking plenty of harmless shocks, I just don't want to take this one.

More Remarks about Capacitors

When you make up your own capacitors, what we have covered will enable you to know pretty well what you are doing. You might, for example, think of borrowing a couple of huge aluminum serving trays with nicely rolled edges, and supporting them parallel, and spaced far enough apart to avoid spark-over. You can easily estimate the capacitance, and the energy stored at a particular voltage.

A word about using glass instead of Plexiglas: go ahead, but keep it clean. Just because it *looks* clean means nothing, in electrostatics. Invisible films of contamination can permit high surface leakage, especially in warm, humid weather.

Glass can be treated to avoid surface leakage. Here is a treatment that didn't work. I hopefully bought a furniture wax in a pressure spray can, made by a famous firm. I applied it to glass, rubbed it as specified, and gave it the warm-breath test. It was no good. (Very probably, if one hunted around among all the many waxes now available, a good one for this purpose could be found.) Next I made just a few little fine shavings of paraffin, dropped them on the glass, squirted "lighter fluid" on them and on a small piece of cloth, and smeared the stuff around. After it had dried somewhat, I removed nearly all of the paraffin, by considerable rubbing. The surface came through fine on the test. You should be able to keep your glass dielectrics in good shape this way. In this test, I charge the horizontal capacitor, having it connected to an electrostatic voltmeter. Two aluminum plates are laid on the tested surface, a couple of inches apart, with each connected to a capacitor plate. The system is charged up to around 30 kV. When my warm breath is blown on the area between the plates, the voltmeter's needle falls at once, if the surface is much affected. If not, it may drop a little, then remain steady.

13. Atmospheric Electrostatics

The more we learn about nature's vast electrical laboratory, the atmosphere, the more fascinating it becomes. Benjamin Franklin, as you know, turned that everlastingly curious mind of his to the clouds, to learn something. He wanted to prove that lightning and electricity were one and the same thing; and did; and luckily, lived to tell about it.

In June, 1752, he sent up a kite in a storm. The kite had a pointed wire at its tip. At Franklin's end of the twine string, he tied a metal key. Then, instead of holding the wet twine himself, he tied a silk ribbon to it, held onto the ribbon, and stood in a doorway to keep the ribbon dry. When a low thundercloud passed over the kite, he saw the wire draw "electrical fire," the fibers of the twine stood out (being charged), and sparks from the key charged a Leyden jar. Good for Franklin! But: much as I urge you to repeat famous experiments, I have to insist that you leave this one alone.

This chapter delves into a few of the major phenomena of atmospheric electricity.

Charged Clouds, and the Electric Field We Live In

The secrets of the atmosphere, electrical and otherwise, are slowly being revealed; and the clear air we look through on a nice day is turning out to be an enormously complex natural laboratory. The sun is throwing charged particles into it; and cosmic rays are smashing up molecules, ionizing some of them, and making the high-up ionized layers that reflect radio waves and make them go farther.

One of the effects is that when you step outdoors on a calm day, you are in a vertical electric field that varies from about a hundred to several hundred volts per meter. And then, when a charged thundercloud comes along, the field may go up to 10,000 volts per meter! Now, why doesn't that voltage kill you? For if *electrodes* from an adequate power source of 10 kV were applied, one to your scalp and one to your feet, you would surely die.

But outdoors, standing on the ground in that electric field, a person, being a grounded conductor, completely changes the field shape. In Figure 32, that "post" represents a child 1 meter high, standing under a thundercloud. The otherwise uniform field is distorted by him, with the field lines drawn solid. The dash lines are equipotential lines. One such equipotential surface, *at ground potential*, envelops him, and there is *no voltage* on him. However: look at that intense field at and near his head. That invites trouble. If a stroke

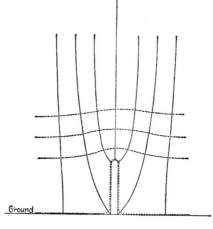

Figure 32. Electric field under thundercloud, concentrated on a post, child, man, or tree.

is developing that "intends" to strike in this area somewhere, and he is the highest object in the area, he may be struck.

When a thundercloud is stabbing away at your area with lightning, "brave" people remain standing. Others go flat to the ground. Also, they know better than to take shelter under lone trees, or high trees in a forest; they know better than to be near wire fences. And they never, if in swimming, remain in the water when there is a chance that a stroke may hit the water.

Lightning

A thundercloud is an extremely active and complex affair. Typically, it is *bipolar*: usually it has a negative charge at the bottom and a positive charge at the top, although this is sometimes reversed. Thus, a giant charge separation has been effected. Just how and why this happens is a matter of intense speculation, argument, and research. There is not now any widely accepted explanation. For half a century, the cloud physics workers have mainly argued that somehow the formation of precipitation causes charge separation (a number of phenomena are involved in this) and that charged raindrops carry charges down. Dr. Bernard Vonnegut, now with the Atmospheric Sciences Research Center at the State University of New York at Albany, has done much work on a quite different theory. Very briefly, it goes like this: A negative cloud base and the field set up between it and ground causes positive point discharges at ground (from tree leaves, etc.), positive ions thus formed being carried up through the cloud center to the top, by *internal updraft*; meanwhile, the positive cloud top attracts negative charges from the atmosphere up there, these ions being carried down to the base by *outer downdrafts.*

Also, the *initiation* of the lightning stroke is somewhat of a mystery. The breakdown of air requires a field intensity of around 30 kV per centimeter. At the ground, the intensity under a low cloud may rise to 10 kV per meter, which is only 100 volts per centimeter. This can initiate the point discharge corona mentioned above, but could not break down the lower air layers and start the stroke. Therefore, it must somehow be started at or in the cloud—and it typically is. When there is anywhere from 25,000,000 to 150,000,000 volts between cloud and ground, a *leader stroke* develops.

The leader stroke, averaging only about 1000 amperes, seems to progress *stepwise* by zigzag corona bursts *in the cloud*, progressing *downward* to the ground in an extremely short time. The *main stroke*, of something like 30,000 amperes (and it may go to ten times that!) at once starts *from* the ground and *goes up*, following the path prepared by the leader stroke. The average energy of a lightning stroke is around 500,000,000 joules!

Going back to that giant capacitor in the previous chapter, with plates 10,000 feet square, spaced 3000 feet apart, and with 100,000,000 volts on them: it is a crude imitation of a low thundercloud and the ground beneath it. The capacitor had 450,000,000 joules stored. That is pretty close to the 500,000,000 joules of average energy of a lightning stroke.

This part of this book happens to have been written at our Canadian summer retreat on an island. From many thrilling experiences, we like to claim that any thundercloud in that area tends to pause over this island, just to show what it can do. I wish you could walk with me for a few hundred yards to see what a cloud's energy did a year or so ago. A tall, lonesome pine, now thoroughly dead, stands on bare rocks some distance from the water. The stroke that killed it had somehow to connect with the water. For the most part, it followed cracks in the rock surface, tossing out sand and such. But near the tree it did a blasting job. It blasted out about six cubic feet of rock, and tossed the big chunks several feet away. Would that have been a good tree under which to seek shelter?

Lightning Rods

The lightning rod, invented by the versatile Benjamin Franklin, is a well-grounded, heavy conductor, with a pointed end sticking up higher than any part of the building it protects. Now: if we tell you not to stand or walk or run in an open field when lightning threatens—because it is true that your chance of being hit is increased—then we should argue that putting up a lightning rod may "attract" lightning. That is, a rod on a barn may be hit more often than the barn would be if the rod wasn't there. However, if the rod is hit, it protects the barn. If the unrodded barn is hit, even only once, it may burn.

Do the rods, on the whole, really do any good, or is this just an old custom perpetuated by lightning rod salesmen? They do help. I began to wonder about this some years ago, and could not think of any ready source of facts. Not at first. Then a hunch came: call an insurance company! And I found that a farmer gets a somewhat reduced rate of insurance when buildings are rod-protected. I don't need to tell you that they do anything like this out of generosity. Insurance companies keep records, for years and years, and they know the score.

Very tall buildings can be targets for lightning. I know that a tower on the campus at Duke University was repeatedly hit in its early days, with the lightning knocking pieces of stonework out of it. Protection by rodding had to be installed.

Ball Lightning

The super-mystery about lightning is ball lightning. Scientists are, after all, human beings; and for many long years, they, almost to a man, scorned the idea that there is such a thing. But no longer. Too many competent observers have witnessed the strange antics of these balls. I myself have met men who have witnessed the phenomenon; and typically, they have had the wits scared out of them by what they saw, and in some cases, heard.

Harold W. Lewis tells, in the *Scientific American* for March, 1963, of a survey of nearly 16,000 employees at the Oak Ridge National Laboratory. Of these, 515 had seen ball lightning. From their descriptions, he writes as follows: "It would appear that the typical lightning ball—or, to use its German name, *kugelblitz*—is a luminous sphere perhaps as bright as a strong fluorescent lamp. The sphere may range in diameter from a few inches up to a few feet, most often from six inches to a foot. It usually materializes immediately after an ordinary lightning stroke. The ball can be almost any color, although green and violet are rare. Most seem to shine steadily, but some pulsate. Normally the ball moves about, sometimes along a conductor and sometimes directly through the air. It can last from a second or less up to several minutes; the median, if one may judge from the estimates of startled observers, is a few seconds. Some balls fade out: others disappear abruptly, occasionally with a loud report. Lightning balls seldom damage anything badly, although they sometimes leave physical evidence of their occurrence. They have scorched wood and burned through wires.

One of my early students, Herbert Curl, had an experience in a summer cabin in 1923. This was at 10 P.M. in late September, on the shore of Bear Lake at North Muskegon, Michigan. He has kindly written to me about it. He was seated in the living room. There was a distant thunderstorm. Something, "not a sound," caused him to look up. He saw a white ball of bright light enter the open front door, coming from a tightly screened porch. It moved slowly across the room, about three feet above the floor, and entered a bedroom through an open door. The telephone there gave one tap. From there, the ball must have entered the bathroom, for the ball chain with which the rubber stopper was supported was found to be broken into a large number of little pieces. The telephone was dead, and its protector block was shorted.

It seems certain that a lightning ball is a *plasma*. A plasma is a region in which both positive and negative charges are present. Plasmas are now undergoing very intensive research, for many reasons, and some of their secrets have been unraveled; but a great deal remains to be learned. At this time, the *kugelblitz* appears to be a complete mystery, both as to how it is formed and as to its subsequent behavior.

Ions Charging Cloud Droplets

Imagine you are wandering in a cloud, surrounded by billions of extremely tiny, freshly formed water droplets. Cosmic rays are busy, turning some of the oxygen molecules into positive and negative ions. As the ions go batting around, some will collide with droplets, be collected by them, and the droplets become charged. Now then, how is this all going to work out? I suspect nearly everyone is surprised when learning the outcome for the first time. Dr. Ross Gunn described it in one of his numerous papers. The table below, in general like one he gives, has a top line of plus and minus numbers, standing for the number of elementary charges on the droplets directly below the numbers. We will start with 32 uncharged droplets:

	−5	−4	−3	−2	−1	0	+1	+2	+3	+4	+5
A						32					
B					16		16				
C				8		16		8			
D			4		12		12		4		
E		2		8		12		8		2	
F	1		5		10		10		5		1

Line A shows the uncharged 32. Next, Line B, 16, *on the average*, would get one plus ion each; and 16, a minus ion each. Now what happens to the minus 16 group? Eight acquire a second minus charge, while the other eight get a plus charge and turn neutral. You are getting onto the game now, and I'll let you have the fun of finishing it. We come out with a large central population with small charges, but on the sides, droplets with higher charges.

If you play this game, starting, say, with 1000 droplets, you get a smoother distribution, and still higher charges at the extremes. Mathematically, a curve showing all this is called a *Gaussian distribution*.

Charges Induced on Raindrops in Clouds

What makes tiny droplets in a cloud come together to form larger droplets? Surely, when oppositely charged droplets find each other, they would join to form larger droplets. But other factors must be at work, too, such as the possibility that a larger drop, falling through droplets, may collide with them and collect them. At this point, I will gracefully step out of this discussion and leave it to the experts.

There is one effect they have pointed out that we should certainly include here. We know that storm clouds can have strong electric fields within them. What might this do to a large raindrop? It would *induce* positive charges on half of it, and negative charges on the other half. Furthermore, it would exert a pull, distorting it away from spherical shape, pulling it out in the direction of the field lines. And if it pulls it in two, there would be two drops, oppositely charged. This may well be another factor in the *very* complex problem of cloud charging and raindrop formation.

Charged Droplets from Bubbles Breaking

In the next chapter, we will talk about charged droplets *falling*, to do their part in making a Kelvin generator perform. But now we will consider the sea, and how it can shoot charged droplets *upward*. This is described in a fascinating book in the Science Study Series, *From Raindrops to Volcanoes*, by Duncan C. Blanchard of the Woods Hole Oceanographic Institution at Woods Hole, Massachusetts.

When the wind makes waves, and the waves break, tiny bubbles in incalculable numbers are formed. These of course rise to the surface of the sea, and break. Only in recent years has it been found that the breaking process has far more to it than you would ever expect.

When the bubble emerges, breaks, and opens out, a tiny jet of water shoots up from the bottom of the cavity, and produces up to five very tiny droplets. And these droplets, for some reason, are *positively charged*. When small enough, these droplets are carried away by the wind, and many are carried up. As they evaporate, the sea salts in them crystallize out. Thus, vast numbers of charges, and salt crystals, get carried up into the atmosphere.

Two very important matters emerge from these new and exciting findings. First, here is a mechanism to account for at least some of the normally positive charge of the atmosphere. Second, the larger salt particles, at rain-cloud level, can account for the formation of *large* raindrops. For, as you no doubt know, cloud droplets, raindrops, and even snow crystals all require *nuclei* of some sort on which to form.

Are Tornadoes Powered by Charges?

A tornado can be tremendously destructive. What starts a tornado, and once started, what keeps it going? A physicist, Vernon Rossow, working at NASA's Ames Research Center in California, suspects that a tornado may be electrically driven. His idea is that when two oppositely charged clouds come near enough together, and there is enough cloud turbulence, streams of oppositely charged raindrops come at each other—not directly but in a whirl. As they whirl, they pull the air into becoming the tornado funnel.

If all this is verified, the important thing is that men's minds have been switched to looking in the right direction—to thinking of how to analyze the case, and making the right observations and tests and models, to get proof. In fact, Rossow has made a laboratory demonstration. He passes a mixture of steam and cool air between two oppositely charged grids. As the steam condenses into droplets, a little tornado, four inches high, develops! This leads him to suggest that if we could manage to shoot a wire through a tornado-in-formation, it could be shorted, and stopped.

Do tornadoes form "up there," where the clouds are? They certainly do, and not at the ground. In fact, on a busy tornado day, a number of funnels are sometimes observed that form, and threaten to extend to the ground, but never make it. At the other extreme a mighty funnel forms, reaches down, and even marches across one or more states, ruining everything in its path.

14. Some More Electrostatic Generators

Late in life when I began to concentrate on electrostatics, the fun of creating new generator forms and new devices got a firm grip on me. For the past five or six years, I have gone to bed nearly every night thinking of the then-present venture, or trying to think of what to dream up next. And then, designing the new equipment, and building as much of it as I had time for, has been enjoyable. Of course, seeing new things perform, or perhaps defy my predictions, and having to change them to make them work—all this is most absorbing. If an oldster can become so intensely interested in electrostatics, why not you? Some of the hundreds of hours I have put in have gone to inventing new forms of generators soon to be described.

Electrostatic generators are not used (with one exception noted below) in industry, for precipitation, separation, electrocoating, and the like. For that, the power supply is a combination of a transformer to get high voltage, solid-state rectifiers to change the AC to unidirectional voltage, and suitable circuitry to make the final voltage as nearly constant as may be needed in a particular application. Such power supplies are far smaller and less costly than would be any *open-air* electrostatic generator it is possible to devise.

The exception is a beautifully developed generator made by a French firm. It is the Sames generator—and I won't tell you how to pronounce that name. A treated glass cylinder is the charge carrier, and it is permanently sealed in a container filled with hydrogen at high pressure. It is widely used in research and high-voltage cable testing. I understand that it is used considerably in European industry.

The Kelvin Generator

For that extremely able and famous man, Lord Kelvin, nothing was too large to investigate, and nothing too small. Likewise for us: we turn from dealing with the tremendous lightning strokes of the previous chapter to the very weak currents of Kelvin's water-dropping generator. Kelvin did not, I believe, investigate lightning. But he did study the atmosphere's electric field, and used water drops to do so. Very probably it was this, together with his use otherwise of induction-type generators, that enabled him to think of this extremely simple and ingenious device. It is shown schematically in Figure 33. Also, the description and Figure 38 in Chapter 16 will tell you how to build your own.

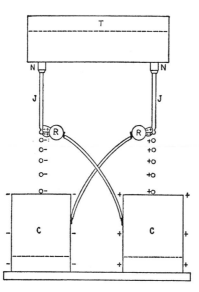

Figure 33. Kelvin water-dropping generator, schematic representation.

There are two tin cans, C and C, placed on insulation. There are two rods, R and R, supported by insulation and *cross-connected* to the cans. Above is a single tank of tap water, with nozzles N and N. The flow is regulated to produce water jets J and J, so that the jets break into drops about opposite to the rods. It is an induction-type generator, like the Dirods; and like them, it is self-starting for the same reasons. If the cans and rods have tiny initial charges and a tiny difference of potential, a plus rod induces minus charges on the end of the nearby jet; and the drops, capturing the charge, fall into the can below and build up its charge. The same holds true, but oppositely, at the other side.

Now, if built as this sketch shows, it would work, but not very well. The jets, at a low build-up of voltage, would be pulled over to the rods and discharge the machine. If, at the left, we then add a plus rod outside of the jet, and likewise at the other side a minus rod, and do so *symmetrically*, two advantages are gained. First, this pull-over would be largely avoided, and second, the *induction of charge* would be *increased*. But again, we can do still better: *surround* the ends of the jets with rings, as Kelvin did.

When a Kelvin builds up, what limits its voltage? Electrostatic forces. The charged drops are *attracted* by the rings and *repelled* by the charged collector cans. My Kelvin, at its limit of 15 kV, puts on a beautiful display of electrostatically curved drop paths. The drops bend around under the rings and spray out sidewise, sometimes spraying the table for a foot or so around.

If the rings are rotated to face not only downward but inward, each spray proceeds to coat the other can.

Some years after building my Kelvin—the first of several—I got around to measuring the rate of drop formation, and was truly surprised to find that each jet makes roughly 150 drops per second.

Many have asked me if there isn't some way to make a Kelvin go higher than about 15 kV. I don't know. The dimensions given in Chapter 16 seem to be "about right." I have not learned of anyone who has juggled these dimensions and obtained a higher voltage. Those who have never seen a Kelvin operate find it hard to believe. And it fascinates everyone who sees it perform.

Next, take a look at a couple of demonstration items that may be used together with the Kelvin.

The Flapper

My flapper, or flexible capacitor, is shown one-third full size in Figure 34. Two thin, springy strips of metal are mounted close together, on a Plexi-

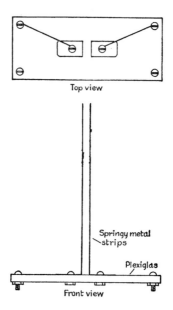

Figure 34. The flapper.

glas base. When connected to the Kelvin, the strips slowly draw together as buildup occurs; then touch and discharge the system; then start all over again.

The Neon Lamp Bank

The neon lamp bank is shown one-third full size in Figure 35. It is also

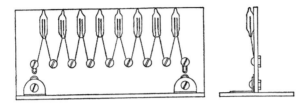

Figure 35. The Ne2 lamp bank. Note the spark gaps at each end, made of pieces of paper clips.

seen in Plate 5. These little Ne2 lamps, as they are called, are quite low in cost. A single lamp fires at around 70 volts. The eight lamps in series should fire at around 560 volts. But if fired with too little energy back of the discharge, the momentary current will be weak, and the flash would be too hard to see.

At each end, I have installed a spark gap, made of pieces of paper clips. These are adjusted to fire at something like 5 kV. This lets the capacitance of the system store up enough energy to give a flash seen all over the room, even in daylight. My Kelvin flashes these lamps about once every three seconds.

You will find, with close observation, that only one of the two little electrodes in a lamp will flash. This is the *negative* electrode. Thus the lamp is a polarity detector.

When the lamp bank's terminals are held near enough to a Dirod's terminals to invite discharges to it, and with the generator going fast, the flashing rate is so high and the light so brilliant that it almost looks to be continuous.

Shake-sphere Generators

Shake-sphere I is a perky little fellow that requires you to shake a couple of spheres, or metal balls, back and forth. Before finding a good name, I called it the Shake-sphere as a joke. Having found no better name, I let it stand. Now that you know the principles of all induction-type generators, you can see why it works, when its parts are identified. You see it one-third full size in Figure 36, and in Plate 5.

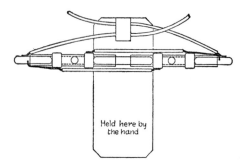

Figure 36. Shake-sphere I generator.

The Plexiglas base, held by hand when you shake it, has two glass tubes epoxy-glued to it. The tubes are just big enough to let two 1/4-inch ball bearing balls play back and forth. The neutral, a 1/4-inch brass rod, is epoxy-glued into the inner ends of the tubes; likewise, for brass terminal rods in the outer ends. Four smoothed-edge aluminum collars are slid onto the tubes and epoxy-glued in place. The connections are clearly shown. The two inner collars are inductors; the outer ones, collectors. You will note that in this design, the charge-carrying balls are charged and discharged, not simultaneously, but alternately—one at a time. After anywhere from 30 to 50 round-trip shakes, the voltage will operate the flapper or the lamps. With nothing connected, the generator goes to about 15 kV.

After losing enough sleep to get the main ideas sorted out, I made my first model—a rough job, using whatever was at hand—in one day. It worked. Then came the "professional" model that you see.

That's Shake-sphere I. Trying to go bigger and better, I went on to Shake-sphere II (Plate 5), with longer chambers and collars, and *eight* balls in each chamber. It is a folded design, with the neutral at the left and both terminals at the right. But I soon found that it does not respond well to *rapid* shaking. Instead, it requires slow operation, being *tilted* back and forth. Apparently, the strings of balls need to travel in close formation to be effective. I believe that the strings of balls should be replaced by aluminum rods of the same length.

A suggestion for simplifying: do without the solid collars—replace them with enough neat wraps of the bared wire itself, to enclose the balls at their end positions.

The Swing Generator

This one is so simple that it is confusing. Plate 5 shows it; so does Figure 37, one-sixth full size. It is built of aluminum and Plexiglas. Aside from

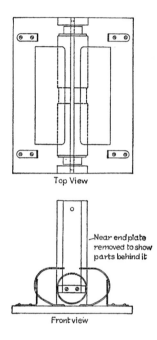

Top View

Front view

Figure 37. Pendulum generator. (Connections not shown.)

screws and the 1/4-inch shaft, there are eight aluminum parts. Four of these are the end-of-swing stops. These are contact pieces. Two are tubes, slid tightly onto the "seat of the swing." The other two are the curved pieces that serve both as inductors and collectors.

The wiring is not shown in the figure, so here it is: First, the upper left and lower right stop pieces are wired together; this is the neutral. Second, the upper right stop piece, which is one terminal, is wired to the collector next to it. Third, the lower left stop piece, the other terminal, is wired to its nearby collector. The stop pieces, made of 1/16-inch stock, are bent so that at the end of the swing travel, the charge carriers are sure to touch *both* stop pieces at one side, then both at the other side.

After preliminary buildup, this generator lights the neon lamps every few swings, and its top voltage is about 15 kV. The easy way to operate it is to pick it up and learn to swing it rapidly.

If you like to dream big, you will sooner or later think of making a giant swinger, adapting a playground swing. Use empty oil drums for the charge carriers. Build the other parts out of salvaged stuff. Even if corona on sharp edges limited the voltage to 15 kV, we ought to store enough energy to make all of the school's teachers jump in unison. Interesting idea? On second thought, figure the capacitance, and the energy stored. And then—don't do it.

More about the Radial Dirods

After Dirod II came into being, I lost plenty of sleep yearning for a Radial Dirod and mentally working out its basic features. At last it took form, and drawings were made. Charlie Hall made the brass hubs, and the Plexiglas parts. I made the other metal parts. In Plate 3 on p. 42, you see the only big Radial in existence at this writing, with its belated offspring, Radial Dirod Junior alongside. The Radial is also seen in Plate 7 on p. 75.

A main object in building the big Radial was to see if corona shields could be eliminated, by using encapsulated inductors. Also, this design easily permits each side to have double inductors, thereby increasing the charge put on the rods. Likewise, the collector plates are double. With the rod going between two plates as the charge is collected, it is virtually in a Faraday cage, and should give up practically all of its charge. It has turned out to be a highly successful machine.

A surprise it handed me is that on being shorted, it reverses.

The Radial Dirod Junior, born last, at the urging of the editor[*] of this book, also reverses when shorted. A guess at the reason for this trait is given in the discussion in Chapter 16 of the Radial Dirod Junior.

Both machines are very smooth-running, since the rods stick out on the plane of the disk.

One very handy feature of the big Radial is in the terminals: they can be swung up or down through a large arc. This change of level is often welcome as a convenience in making connection to various pieces of demonstration equipment.

Dream Up Your Own!

By now, you are thoroughly familiar with the basic ideas of charge induction, charge separation, charge collection, and voltage buildup; also with a number of general features of construction, and with variations of these fea-

---

[*]Of the first edition of the book.

tures. When you get into Chapter 16, you will find a lot of Dirod details to complete your knowledge of the family.

Now go ahead and invent your own new form of generator!

It will have to make use of the same principles, of course. But its form, and design, and use of materials may be quite different from anything made before. *Humans are creative*! There is nothing quite like the satisfaction that comes from creating something new.

15. On This and That

What is electrostatics? Here we approach the end of the book, and I haven't even defined that term. We like to think of a good definition as a concise little group of words that includes whatever belongs, and excludes that which doesn't belong.

One writer has stated that electrostatic fields are fields that do not change in time. If we accept that (which I do not), it wipes out most of the content of this book. When an electrostatic generator is building up, its electric field is changing, but it is still called an *electrostatic* generator. In the three great areas of precipitation, separation, and electrocoating, the applied voltage is reasonably constant in some cases, but in many it is far from constant. Yet these applications are universally classified under the heading of electrostatics.

So it comes to this: I suspect that as of now at least, it would be impossible to frame a short definition of electrostatics that would meet with wide acceptance. For my part, I refuse to attempt a definition, and will simply claim that it is what this book is about.

Why Dust Sticks

If dust particles simply drifted to a wall and happened to settle there, a slight puff of air should make them come loose and start them swirling through the air. But it isn't that easy to clean a wall. The dust *sticks*. Nearly all fine particles have at least some charge of the one kind or the other. Two different kinds of particles can charge each other, by touching, or being rubbed together. Picking up ions from the air is another way for dust to get charged. There might be other ways.

When a dust particle charged positively, for instance, drifts close to the wall, negative charges are induced on the wall surface, an electric field is set up, and the particle is pulled to the wall and stuck there electrostatically.

This is often called the *image effect*. Imagine that the wall is taken away, and that there is an identical particle, oppositely charged, as far inside of the plane of the wall surface as the real particle is outside of it. The outer hall of the electric field would be identical with that which you find between the real particle and the real wall. There may be other reasons for dust sometimes sticking so tightly, but electrostatics certainly is a major factor.

Are Negative Ions Good for Us?

The air we breathe and live in has a highly variable content of ions, both as to total number and as to the division between plus and minus ions. This is true outdoors. Indoors, the ion count is often greatly reduced by tobacco smoke, and by ions going to various somewhat charged surfaces. Any reasonable medical man would have to admit that the ion content *possibly* has physiological effects; he would have to admit the further *possibility* that deliberate control of it, for daily living, might be beneficial, and the still further possibility that treatment of some patients by having them inhale air somewhat highly charged with one kind of ion or the other might help—or that wound healing, say, might be benefited.

The argument about this has gone on for half a century or more. Rather recently, a fair amount of research has been done. The situation at this writing is that while one side insists that suitable *negative* ion concentrations are quite beneficial, the other side insists that the case is not proved. So: if you like to get in on a red-hot argument and help to settle it, one way or the other, here is your chance. Don't worry about getting there too late, for this research may go on for quite some time. Learn all you can about it, prepare to enter it with an open mind, and then help to determine the facts.

Ions in Liquids and Solids

Throughout this book we have considered ions as charged atoms or molecules in a *gas*, such as air. But there is another kind of ion. Consider a molecule of common salt: NaCl, or sodium chloride. There is a bond that holds these atoms together. When you dissolve this salt in water, the water molecules have the ability to break the bond. Also, instead of *neutral* Na and Cl atoms wandering about, we know that the sodium has lost an electron, becoming a positive ion; and the chlorine has gained an electron, becoming a negative ion. The solution as a whole remains neutral, of course.

If you do as Faraday did, and conduct a large number of experiments by passing a current through solutions and observing the effects, you will find—as he did—that the weak electric field set up between your electrodes makes these two kinds of ions drift in opposite directions. Their movements constitute the current in the electrolyte.

Faraday, wise man that he was, called on a friend who was a language scholar, for help in giving these little things the right kind of a name. That is how he came to call them ions—for *ion* means *traveler*.

Now we turn back to that little aluminum cube (or any other solid conductor) and its electron cloud, and its vibrating but otherwise pretty well fixed atoms—and the fact that these atoms are positively charged. Should they be called ions? If, for the most part, they are fixed, and do not travel, we would hesitate to call them ions. Even so, enough respected writers have called them ions to make it seem that the practice may be generally adopted. Furthermore, it is known that some of the atoms in solids *do* migrate, and these, at least, are travelers.

The Electret

Pass for a moment to magnetism. We can magnetize a bit of hard steel and make a compass needle. Hard steel magnets tend to "age," meaning that they slowly get weaker. So then came an ingenious Japanese who made an alloy of aluminum, nickel, and cobalt, called alnico. This can be magnetized strongly, and it is highly permanent. Such permanent magnets have long had many important uses, and their applications are rapidly increasing.

Long since, men have dreamed of an analogy. Corresponding to the permanent *magnet*, could we not have a permanent *electrostatic*? Or more briefly, an *electret*? In a sense, we do have any number of *electrostats*, but we call them capacitors. A good capacitor can hold a charge for a long time; and if you merely require it to furnish *voltage* (not current) you do have a "permanent' electrostat, if "permanent" means for a day or an hour. But that isn't the dreamed-of electret, which would be something we could charge, have retain its charge permanently, and have it stand ready to deliver voltage (not current) to maintain a needed electric field.

There are many waxes. One of them is *carnauba* wax. Melt it in an insulating container, with flat electrodes on top and bottom. Apply, say, 10 kV, and maintain it for as long as two days, while we see to it that the wax cools *very* slowly. This makes an electret. It is *relatively* permanent. It may retain most of its voltage for months, if properly handled and kept. The vexing thing is that it is not really permanent, and is by no means understood. Some good research teams have worked at it, using various wax combinations; and from what I hear, I suspect that many competent teams have tried to come through with the electret. All have failed. When—and if—it does come through, many uses will promptly be found for it.

(Since the above was written, the Northern Electric Company Limited of Ottawa, Canada, has announced the development, by Peter Katovic and others, of an electret using K-1 Polycarbonate, to be used in telephone microphones. Tests indicate a very long life. This application will undoubtedly be followed by others.)

The Electrostatic Speaker

The speakers in radio and TV sets are driven electromagnetically. Now, the electrostatic speaker is not only a possibility — it was invented years ago. Since the diaphragm membrane is very thin, its inertia is very low, and the speaker can offer very high fidelity. It must have a "biasing" voltage of one or two kV (very constant DC). Thus, if the electret is ever made to be completely dependable, it could nicely serve as this voltage source.

To be truly "flat" in frequency response, the vibrating membrane should be part of a spherical surface in shape. This presents a difficult problem. Another requirement for true fidelity is that, for the very lowest notes in full orchestral bass, the speaker has to be 8 or 10 feet wide and high. Rather large! Nevertheless, new approaches to solving these problems may be thought of, and someday, electrostatic speakers may be in common use.

The Most Remarkable Capacitors Are Inside of You

If you study biology—and everyone should know something about it—you will learn something about cell structure and function. The more that is learned about it, the more difficult it is to believe that such a complex thing as a cell could have evolved, and that it can carry out the many things it does to maintain life.

Now, every living cell, plant, or animal has the ability to keep its soupy interior *lower* in potential than the tissue or the blood outside of it—roughly 70 or 80 millivolts, in fact. And it maintains this minus 70 or so millivolts, by virtue of having a *membrane* surrounding it, that "knows" how to juggle sodium, potassium, chloride, and other ions of the plus or minus kinds.

Next, all cells except two kinds are what you could call electrically static: they just maintain that difference of potential. But *nerve* cells and *muscle* cells are radically different—different in shape, and vastly different in what they can do.

In dissecting an animal, the whitish thing you call a nerve, is really a *nerve trunk*: a bundle of from several to a large number of nerve fibers running along together—each coming from and going to the right places, and standing ready to transmit the right messages. A *neuron* is a complete *nerve cell*. In many nerves, it has a *cell body* at one end, with a thin thread, or nerve *fiber*, called an axon, going somewhere from it. A bundle of axons makes up that nerve trunk we commonly call a nerve. And the entire neuron is covered with this remarkable membrane. It is a very thin membrane, less than 100 microns thick (100×10^{-4} centimeters).

This membrane can, and does, somehow maintain a low concentration of sodium ions inside compared to outside, and maintains imbalances in the other kinds of ions. How it manages all this is now the subject of very intensive research. The entire effect, in the nerve *at rest*, is to maintain that minus 70 millivolts—and wait for action!

Before taking up the action, note this amazing fact: that the field through this membrane has an intensity of the order of 100 kV per centimeter! And note this too: that membrane is a *capacitor*.

Now comes the action. A nerve message, or messages, coming from elsewhere to the cell body, causes part of its membrane to become depolarized. Electrically, it breaks down. We say it is being "fired." This effect spreads very rapidly along the axon. To help get the picture, think of stringing up a very long clothesline, placing your fingers around one end of it, and, running along it, slide your closed hand along it. Your hand-enclosed region is, at any instant and position, the *zone of firing*. There is a *sodium inrush* briefly where your hand is, and overlapping that, a brief *potassium outrush*. Ahead of your hand, firing is yet to occur, but it is being prepared for. Behind your hand, *restoration* to the previous resting state is taking place.

In nerves going to the muscles we use for *voluntary* acts, the firing message is speeded along at from 5 to 100 meters per second, depending partly on size of fiber. Also, the time for restoration is so brief, that as many as 50 of these impulses per second may arrive at the muscle fiber being stimulated.

Next, the muscle. It is not just a hunk of meat, somehow able to shorten as a rubbery whole to move an arm. It is a highly organized affair. A muscle is a small or large bundle of relatively long, tiny *muscle fibers*, several of which get the message from a single nerve ending. These cells, or fibers, have their own remarkable membrane, maintaining that negative potential in the resting state. When the nerve stimulus arrives, it fires the membrane of the fiber, and the firing zone runs along it, just as it did in the nerve fiber. This, in turn, causes the complex arrangement of molecules in the fiber to make a contraction. How this all happens is only partly worked out. You can read further along these lines in *Nerves and Muscles* by Robert Galambos.

Nature has a way of taking an "idea" and using it in as many ways as possible. If a muscle cell in long fiber form can develop nearly a tenth of a volt, why not evolve flat, plate-like cells, stack them in series, and give out a shock? That is precisely what evolution has done for the electric eel and some other fishes. By turning loose the energy stored in these living capacitors, these animals can shock their prey and stun it, for capture. Not only that: in the muddy waters of the Nile, in which no fish can see to get about, *Gymnarchus niloticus* has developed a battery of these specialized cells for setting up an electric field in the water. When the field is changed by any nearby object, sensory organs in the skin of the fish do a perfectly amazing job of telling the fish how far away it is, and what it may be.

Thus, the most amazing capacitors came out of evolution, and you have them by the billions. In your brain alone, you have about twenty billion neurons. Put them to good use!

Some New Electrostatic Developments

Electrostatics is a very old but a greatly neglected field. As more workers become interested in it, and as more needs develop that cannot otherwise be met, electrostatic applications are bound to come along. They are coming now. Let us take a look at some new and radically different developments. I hope they stimulate your imagination.

Printing. Printing a picture or a name or a trademark on a raw egg—did you ever dream of doing that? It can be done. This new process uses a fine screen, a special powder, and an electric field. I have not been able to learn other details. This new form of electrostatic printing can put pictures and labels and other impressions on rough or uneven surfaces.

I have already mentioned the Videograph in Chapter 10, for high-speed printing of address labels. This process has now been adapted to replace the typewriter, which has become too slow to print out the output of some of the faster computers. The process, I read, can electrostatically print 120 characters per second.

Pictures. General Electric has announced a new picture-making process. A new kind of film is used, such that positive charges can be sprayed on the top surface. Negative charges then appear on the bottom, and together these try to squeeze the film. Next, exposed to light in a camera, a photoconductive effect removes the charge where light strikes. The film is then warmed in an accurate way. Being a thermoplastic, the remaining charges do squeeze the film, making it thinner where they are—and the image is permanently recorded. This, I believe, is still in development.

Surgery. A problem in surgery is adhesions: after surgery, parts may grow together that should not thus be joined. The right kind of a barrier might prevent it. Here is where Dr. John P. Gallagher of Georgetown University Hospital has made a contribution. Gold is tolerated indefinitely in our tissues. In the extremely thin form of gold leaf, it can be used as a barrier—provided, first, that it could be picked up and applied; and second, that it would consent to snug down tight onto any and all irregularities. Gold leaf is so very delicate that no ordinary means of picking it up will do anything but ruin it.

Dr. Gallagher first electrifies a camel's hair brush, then picks up the piece electrostatically. When the piece is touched to the tissue, it immediately snugs down tight. He also uses it to shut off minor but persistent bleeding. It has been stated that since the leaf is charged, it is electrostatic attraction between leaf and tissue that makes it snug down as it does. I have disagreed with that. Certainly, there is attraction, *before* touching. But the instant the piece of leaf touches wet tissue, the charge would be drained off. I maintain that capillary attraction is what snugs it down, just as a wet piece of paper clings to your hand. Whatever the reasons, it works, and it is an admirable contribution that uses—please note—a very old material: gold leaf.

Electrostatic Gun. Could there be an electrostatic gun? Yes. A model of one, called a Hypervelocity Gun, is on display in the hall near my door. Harold C. Early (mentioned as having given me my first conducting rubber for Dirod brushes) is one of our most ingenious research men here at the University. He and his group developed the gun. It is a blocky little affair you could hold in your cupped hands. It is arranged so that the discharge from a capacitor vaporizes a bit of foil in the gun, doing it just behind a film of that tough plastic, Mylar. Whereupon, the extremely high pressure punches a little disk out of the Mylar, drives it along a short barrel, and sends it, in a vacuum, ahead at 6 *miles per second.* It smashes as far as 1/4 inch into a lead target, and knocks some lead out of the other side.

Why all this? Well, in this Space Age, the effect of meteoroid impacts on space satellites and vehicles must be looked into. This is one way of testing such impacts. Take a look at Harold Early's figures. That disk of Mylar is only 0.01 inch thick. The aluminum foil, vaporized behind it, is 1 square centimeter in area and only 0.0007 inch thick. To vaporize it, Early uses the discharge from a 70 microfarad capacitor, charged to 20 kV. And the extremely short current rises to a maximum of 400,000 amperes! This isn't a test that is popped off every few minutes. The gun is completely destroyed. Next test, another gun. By the way, using capacitor discharges is old stuff to Harold Early. Some years back, he invented the method for using such energy for the high-speed punching of paper.

Space Fuel. In a space vehicle, a thruster must be turned on, burning fuel from a tank. At one end of the tank is a vapor vent. At the other, is the liquid fuel connection. But out there, in zero gravity, the liquid may be loafing around anywhere at all in that tank. This has been quite a problem in space engineering. An electrostatic way of solving the problem has been described by J. M. Reynolds, Manager of the Advanced Systems Department of Dynatech Corporation. At the liquid exit end, place some plates (or even screen) making them fan somewhat outward from that end. Apply voltage, setting up an electric field between adjacent pairs of plates. These are *nonuniform* fields, and the fuel will be drawn toward where these fields are more intense. The forces are weak, but so what? With no gravity present, weak forces will turn the trick.

Plants in Electric Fields. A young researcher, Larry E. Murr (see bibliography) has been doing research on the effect of electric fields on growing plants such as orchard grass, sweet corn, sorghum, and beans. Different plants react differently. When the field intensity is high enough to damage the leaf edges (no doubt due to corona formed there), the damage causes some plants to grow faster. Others are retarded. It will be interesting to follow Murr's work as it is continued. Young scientists headed for biology might consider working up Science Fair projects along these lines.

Corona Chemistry. The history of science is full of instances in which someone reaches back into dusty volumes for some curious but "useless" effect, and, with the help of new knowledge and new needs, brings the effect to life and into usefulness. Such is corona chemistry. It has long been known that certain chemical reactions could be effected in corona. The subject has now become one of very active interest.

Ink Jet Oscilloscope. Oscilloscopes are universally used to record signal traces on photographic paper or film. Also, great use is made of oscilloscope recording devices at lower frequencies, in which a pen or the equivalent directly records a trace on moving paper.

Another way to record a signal trace is described by Richard G. Sweet (see bibliography). The trace consists of electrostatically controlled ink dots. An extremely fine jet of fountain pen ink is broken into precisely uniform droplets, by vibrating the nozzle. This makes as many as 100,000 droplets per second! The droplets are individually charged by a nearby inductor (just as in the Kelvin generator) so that the charge is proportional to the signal strength at that instant. Even so, this alone would merely draw a straight line of dots on the passing paper. But next, for most of the way from jet to paper, the droplets pass across a uniform electric field between two plates. This deflects them in proportion to the charge carried. It is interesting to note that for this very modern device, some of the ideas used date back to 1833, 1859, and 1867.

The Future of Electrostatics

Electrostatics offers not just one or two phenomena, but a whole group. And they can occur in combinations that are strange, mystifying—and useful.

Since precipitation, separation, and electrocoating have been around long enough to be utilized in large industrial enterprises, does it mean that these applications have already reached their peaks of development? It means no such thing. As more theoretical understanding is reached, and as ingenious men think of more things to do and better ways of doing old things, each field is certain to grow.

The previous section covered a number of newer electrostatic developments. I have often wondered, while writing this book, what still newer developments will be announced shortly, that I wish could be mentioned in it. And who will be making new discoveries and new applications in the future? Young scientists who get started on electrostatics in high school, who carry their interest through college, will, I fully believe, be a long way ahead of others in the chance of making these contributions.

I would especially welcome the *experimentalist* into the whole field of electrostatics applications. The highly trained theorist who knows all of the principles, all of the available theory, and all of the mathematics needed, can make contributions—as long as things behave themselves in sufficiently simple fashion. But time and again, electrostatics won't behave itself. The shapes of electric fields can be so complex as to defy analysis. Corona phenomena can get out of hand and refuse to fit those beautiful equations. This is where the experimentalist—such as the great Faraday was—can save the day. He does not work blindly. He also informs himself about principles and effects and phenomena of the several kinds. But he is willing to *try* things. Like Faraday, he is ingenious. Like Faraday, he can think up new attacks. Like Faraday, he can build apparatus himself. Like Faraday, he is a keen observer, always looking for the expected, but especially observant of the unexpected. Like Faraday, he gets a *physical understanding* of what goes on, as nearly as that can be managed. Like Faraday, he lives with his problem day and night. Like Faraday, he never gives up. And every so often, like Faraday, he gets the thrill of succeeding where all others have failed. We need more Faradays!

And remember this, as we close this chapter: this book deals only with what I call open-air electrostatics. Think of all the other things electrostatics can do in controlled atmospheres, in partial or nearly complete vacuums, and so on.

I hope you have enjoyed reading this book as much as I have enjoyed writing it. And I can wish nothing better for you, than that you create your own electrostatics laboratory and have a world of enjoyment.

16. Some Final Hints

Corona Phenomena in More Detail

There is a profound difference between positive and negative corona. In negative corona, multitudes of electrons leave the electrode surface. Let us follow one of them. It collides with an oxygen molecule, knocks out an electron, forms an electron-positive ion pair. Now there are two electrons, ready to form two more pairs. This becomes an avalanche. The ions are urged to the electrode, give up their charges, and hit hard enough to knock out more electrons. Meanwhile many electrons form attachments, making negative ions that are urged along the field *away* from the electrode. In the process, many oxygen molecules become excited, with electrons raised to a *higher energy level*. As these fall back to a lower level, *visible light* is emitted with which the corona is *seen*. Also, *ultraviolet energy* (invisible) is emitted. Some of this, striking the electrode, causes more electrons to be emitted. A prime factor to note is that action is initiated at the electrode surface, with an electron making an avalanche. Hence the spotty, brushy character of the corona. It tends to have streamers, to be irregular.

In contrast, positive corona does not initiate at the electrode. It starts out at the edge of the visible corona. From cosmic rays and so on, there always are some loose electrons. Follow a loose electron outside of the corona, speeding toward it, hitting an oxygen or nitrogen molecule, and forming an electron-positive ion pair. The ion starts its journey back toward the negative electrode. Perhaps one of the *two* electrons we now have forms one more pair, while the other drifts to the electrode. This corona also has excitation of molecules and emission of visible and ultraviolet energy. But since this electrode is a receiver and holder of electrons, the ultraviolet cannot make it emit electrons. Since the action begins at the outer corona edge, positive corona tends to be a uniform glow.

The two coronas are also different in their effects on whatever they are used for. In precipitators it is often true that the voltage used for negative corona can be twice as high, without spark-over, as with positive corona. Negative corona is used in most corona-dependent apparatus. An exception: the Westinghouse Precipitron uses positive corona—for that design, it makes less ozone.

There is far, far more to corona phenomena than what is said in the above brief sketch. The study of corona by theoretical and physical research has kept many workers busy for years, and the answers are not all in yet, by any means.

Sphere Gap Data

In case you happen upon a pair of good spheres, but not of a size given in the table below: plot the data in the table as a family of curves; use your judgment to draw the curve for your size, in among the family. It should pretty well be the curve you need.

Table 4. Approximate Spherical Gap Data
Gap cm given in body of the table

	Sphere diameter, cm				
kV	*2.5*	*3*	*4*	*5*	*10*
10	0.30	0.30	0.30	0.30	0.30
20	0.61	0.61	0.61	0.61	0.61
30	0.95	0.95	0.95	0.95	0.95
40	1.40	1.32	1.30	1.30	1.30
50	2.00	1.82	1.73	1.71	1.65
60	2.81	2.40	2.21	2.16	2.02
70	4.05	3.16	2.80	2.68	2.41
80		4.40	3.50	3.26	2.82
90			4.40	3.93	3.28
100				4.76	3.75

Field Shape Indicator Details

Two indicators are in Plate 7 on p. 75; the drawing is Figure 5. At the left (Plate 7 on p. 75), paper strips were cut long enough to form loops, with ends overlapped and glued. These were squeezed flat, but with a bulge between sides. Pierced with a needle larger than the pin, they turn freely on the pins. Pins have plastic spherical heads: this keeps friction down. A round bead behind the flattened loops, glued on, holds the loops out from the Masonite panel. At the right, Plate 7 on p. 75, the indicators are cut from 3- by 5-inch cards and mounted on cylindrical glass beads; as above, with backing beads to hold them out from the plastic sponge into which the pins are stuck.

Reversals: a very striking effect. The Radial Dirods reverse when short-ed. Operating the (left-hand) paper loop indicator, when generator polarity reverses, the indicators flip and reverse. They become polarized. The (right-hand) piece-of-card indicators want to reverse but seldom manage to do it.

Plans for the Kelvin Generator

The essential parts are in Figure 38, one-sixth full scale. The framework

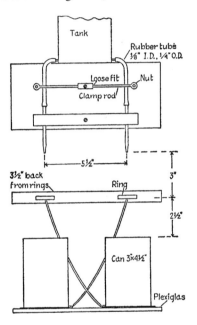

Figure 38. Kelvin generator.

of wood to hold these parts is not shown. A rod, clamping onto a rubber tube, regulates the jet. The clamp rod has a washer soldered edgewise to it, with a nut on a screw to push the rod against tube.

The nozzles are glass tubes. A single tube, flame-softened, is pulled out and broken off to make both nozzles. The nozzle tip hole is about 1/16 inch. Another way: use straight tubes, fill lower end with epoxy, drill the epoxy.

The rings are cut from brass tubing 1 inch in diameter, 1/16-inch thick, edges smoothed (various forms of rings or collars will work, such as collars 1 inch high, of gauze or screen). The rings are supported by rods stuck into holes in Plexiglas piece further back. Dimensions need not be rigidly held to. Recommendation: stay fairly close to those given.

The aluminum plates rest on a Plexiglas base, these being cross-connected to the rings by rubber-covered wire. This "free can design" lets the cans be emptied back into the tank without disturbing connections. The connecting wires hook onto the rods that support the rings, back near the support; if allowed close to the rings, they distort the field and cripple the action. The plates extend out in front of the cans, ready to accept the lamp bank or the flapping capacitor.

The plates, I have found, should not rest directly on the Plexiglas base, for this invites too much surface leakage. Elevate them by perhaps one-quarter inch by cementing four tiny insulating feet to the under side of each.

Many youngsters have written to me for these plans. It has been most rewarding to hear from them later, telling of winning ribbons in Science Fairs.

Materials and Methods

Young experimenters need to learn all they can about how to make things, what to make them of, and so on. This is true whether it be for electrostatics work or for preparing any top-notch exhibit for a Science Fair. Or how about just satisfying yourself by doing a really good piece of work? Later, as scientist or engineer, you will find that the more you know about how things are *properly* made, the better off you will be.

Machinists

Mechanics, machinists, instrument makers—one of these can be found in almost any community. Find one, make a friend. You may need his advice. He may find a few screws for you, to save buying a whole boxful. He may lend you things—drills and taps and dies. We need friends at any stage in life. A man making his living with machine tools is often glad to help an enthusiastic youngster. You can learn much from him. I am still learning from my machinist friends.

Plexiglas

An excellent insulator material. Surface leakage is very low, even in humid weather. Methods for working on it are treated in detail later (Dirod Junior instructions).

Glass

Suitable for corona shields. It can be used instead of Plexiglas for dielectric in the horizontal and vertical capacitors. The surface is easily contaminated and can be very leaky; it can become useless in high humidity, unless treated, in capacitor service. One treatment: dissolve a few very small paraffin shavings on the glass, in lighter fluid. Rub with cloth to remove all but a thin film. Another method: spray with two or three coats of Krylon.

Wood

Wood, Masonite, cardboard, paper—all can be quite good insulators when quite dry; very poor to useless, when not.

Paraffined wood

In my opinion, this has fine possibilities for replacing a lot of the more expensive Plexiglas shown in this book. It is strong and light. Holes drilled in it are clean and sharp. In general, shape the wood piece before treating it, but drill afterward. Dry it thoroughly with warm, dry heat. Weight it down in the paraffin bath for an hour for little pieces: perhaps eight hours for large ones. Not recommended for this: plywood. The glue might be electrically leaky; and it might prevent complete paraffin penetration.

Paraffin

Soft and weak by itself, but an excellent insulator. Warning: *never* melt paraffin directly on gas or electric or other heat source—high fire hazard. Always have the paraffin pan in hot water, to melt, or maintain paraffin bath. If parts are to be epoxy-glued to paraffined wood, glue them on *before* the paraffin treatment. I would not expect epoxy to stick to paraffin.

Plastic Sponge

The Plastic Age has much favored us. Plastic sponge material can be had at the dime store as kneeling pads, 2 by 8-3/4 by 14 inches. Make fine insulative slabs for various demonstration items. One is shown, Plate 7 on p. 75, right-pins of the field shape indicator are stuck into it. White Styrofoam slabs (stiff) are available, and very useful.

Krylon

An insulative coating coming in various colors in pressure spray cans. It is much used to shut off bothersome leakage over surfaces, and is also used to reduce or stop corona.

Corona Dope

This is the actual name for it. TV repair shops have it, or can tell where to order it. It is dark red, thick, sticky, and is used for shutting off corona. It may be seen in Plate 2 on p. 41, on outer ends of rods. The rotor was clamped with the axle vertical, rod ends down. Using a small cap from a bottle, or such, you fill it with the dope and bring it *up* to coat a rod end. After drying for half a day, another coat is applied.

Tapes

Many adhesive tapes are available. They are indispensable for experimental work in electrostatics, and can be used many times for holding things on, or together, temporarily. They are used permanently in Dirod Junior, for holding the brushes.

Soldering aluminum

Don't. Leave it to the experts.

Epoxy glue

No experimenter can afford to be unaware of what epoxy can do. Epoxy appears in many formulations. We deal here with the glue, now widely available. It comes in two tubes. It can be had in black, white, or clear. Equal parts must be *thoroughly* mixed before use. No need to hurry—there is ample time before it begins to set. Quite strong setting may take twelve hours. Setting in about two hours can be had, by using an infrared lamp to heat the parts hot enough to be uncomfortable to the touch. Surfaces joined must be cleaned.

The fresh glue will *not* usually help to hold parts together. Often, you must rig up some other, temporary method of holding the parts together. The choice of *applicator* can be very important. A needle can nicely pick up very small amounts, and place them accurately. Don't hesitate to make up your applicator of the right size and shape for the job. Epoxies are insulators. Epoxy may creep between two metal parts, preventing electrical contact, unless they are held tightly together while setting occurs. A bit of silver paint applied over the joint makes sure of contact. TV repair shops use this paint: go to one and get permission to use the very small amount needed from their in-use bottle. Epoxy's shortcoming is that it is brittle. Drop something, and the glued joint may break. It is easily reglued.

To epoxy onto paraffined wood, epoxy before paraffining.

Needles for little shafts

As in the Franklin motor, a piece of a needle makes the shaft. Needles are too hard to file. Clamp the needle tightly in a vise; place some scrap metal against it, resting on the vise, so that the coming impact hits close to where the needle is held. Hit it a sharp blow. Safety note: the broken piece flies away fast—it could ruin an eye.

Rod material

If standard rod stock for Dirod rods is not available, go to the hardware store's kitchen supplies, look for "potato nails," "heat rods," or whatever they may be called by a maker. These are stuck into big potatoes, and roasts, to conduct heat into them, and are made of aluminum. They have one pointed end; the other is curled, or has a head. The degree of smoothness will depend on the maker; some may need smoothing. Of course, ordinary large steel nails could be used, except that it would take a lot of work to round and smooth the ends.

Spheres

Smooth, accurate spheres are expensive. One-inch, $2.85; two-inch, $5.95—quoted by Universal Voltronics Corp., 17 South Lexington, White Plains, N.Y. Another source: Industrial Tectonics Inc., 3686 Jackson Road, Ann Arbor, Mich.

Abrasives: Metal smoothing, polishing.

Emery cloth: the name familiar to all who smooth metal; the old-time member of the *coated abrasives* family. Long carried in hardware stores, emery is a natural composition of corundum and iron oxide. It is rapidly being replaced by *aluminum oxide* cloth (one trade name, for example, is Aloxite), which does a better job. For rough fast rounding of an edge, use Grit 240. Follow with the finer grade, Grit 320.

For fast edge-rounding and smoothing, use the cloth in strip form. Clamp the plate in a vise, take an end of the strip in each hand, pull back and forth. A worn strip of 320 gives a finer finish than fresh cloth.

To polish to a gleaming surface: go on to fine steel wool, or to crocus cloth, which, being iron oxide, is red.

Be on the lookout

Never go into the dime store, hardware store, auto supply store, bike shop, or store where model cars and planes are sold, without looking for materials and parts you can use or adapt. An experimenter sometimes finds exactly what he needs in a most unlikely place.

More about the Dirod Family

Often there is more than one way to construct a feature, and get the same result. This is brought out by studying the whole family—making it more feasible for you to select a generator to build, with materials and methods suited to your situation.

Comparing the two Juniors: Dirod Junior is more compact; Radial Dirod Junior is simpler, with a little higher voltage.

Dirod Junior can be "stretched," with rods 8 inches long instead of 4 inches; other parts should be lengthened axially to correspond. This should about double the current output at the same speed and voltage.

Dirod I and Dirod II can be stretched. But these are heavy rotors, and they should then be *between* bearings instead of overhanging. Move the minor end plate to front, as in Dirod Junior; also, have the longer rods extend equally each way from the disk, as in Dirod Junior. Also, the neutral and brushes could be brought to the front in a like manner.

The big Radial Dirod, largest of the family and the most spectacular, has the highest output.

There follow complete plans and instructions for Dirod Junior; next, the same for Radial Dirod Junior; and finally, the Dirod Family Data Table.

Building Dirod Junior

Remarks

If you have never built anything as ambitious as a Dirod Junior, you will get invaluable experience with the nature of materials, in adapting them to your needs, and in learning how to use hand methods as well as simple machine processes to carry your project through. The experience and fun you get out of it will repay you for all the energy and money you invest. And you will have an endlessly fascinating electrostatics generator.

If you prefer to build Radial Dirod Junior, read this first; by giving full instructions here, those for the Radial Junior can be compressed.

Drawings

Front and side views are shown about 1/5 full size, Figure 39; and parts, Figure 40. Plate 4 shows the machine. You will do well to make full-size drawings before proceeding; it helps to avoid mistakes. Note that in the side view, parts have been left off, to enable other parts to be seen; this view also shows only two of the twenty-four rods.

Some main parts

A sub-base of plywood supports both generator and a motor behind to drive it (motor not shown in Plate 4). A wood base about an inch thick is screwed to the sub-base. The two main vertical parts attached to base are end plates of Plexiglas—minor end plate in front, main end plate in rear. The minor end plate supports neutral brushes. The main end plate supports round collector plates, attached by screws. Each collector supports its own inductor. The main end plate supports corona shields, each attached by a single screw. Rotor end play and rotor position are controlled by round Plexiglas pieces on the shaft.

Dimensions

They need not be copied precisely. The figures are too small to show dimensions, so they are given here. Base, 1 by 4-1/4 by 5-3/8 inches. Main end

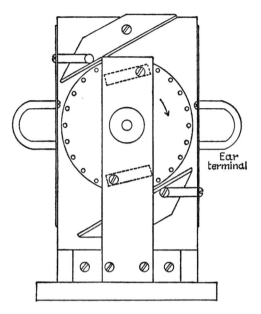

Figure 39a. Dirod Junior, front view.

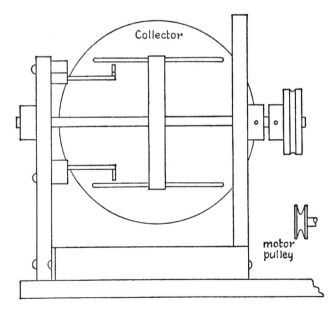

Figure 39b. Dirod Junior, side view.

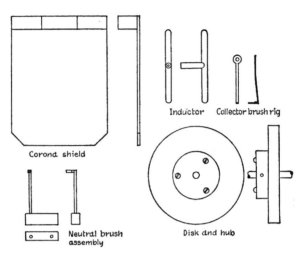

Figure 40. Dirod Junior parts.

plate, 1/2 by 4-1/4 by 8 inches. Minor end plate, 1/2 by 1-1/2 by 6-3/4 inches. Shaft center 2 inches down from top of minor end plate. Shaft, 3/8 inch; could be 1/4 inch. Plexiglas disk, 1/2 inch thick by 4 inches diameter. Brass rods, 24 in number, are 1/8 by 4 inches, extending equally each way from disk. Disk center plane, 2-1/4 inches from main end plate; this brings rod ends 1/4 inch from that plate. Aluminum collector plates, 6 inches in diameter, 0.042 inch thick. A line joining collector centers passes through disk center. Attachment screws for collector plate, 1-1/2 inches apart. Ears (terminals) are 1/4-inch copper tubing bent to inside radius of 1/2 inch or more. Shaft length, 9 inches. Pulley made to speed reduction around 4:1.

Parts, Figure 40. Corona shield, 1/8-inch window glass or Plexiglas, 4 by 5-1/4 inches; epoxy-glued to its Plexiglas support, 1/2 inch thick, and 3/4 by 4 inches before corners were beveled off. The attachment screw is on vertical centerline, 3/8 inch down from top of end plate. Shields are sloped 30 degrees. One inductor is shown, brass or aluminum, 1/4 inch in diameter. The inductor is 3 inches long, attached to support rod of same material 1 inch overall length. The end of the support rod is hollowed out 1/8 inch to fit to inductor. Epoxy joint; epoxy bridged with silver paint. Center of top inductor is 1-1/4 inches down from top of main end plate, thus fixing screw hole in collector.

The disk is shown screwed to a Plexiglas hub 1/2 by 2 inches in diameter. The hub, if needed, is forced onto shaft, or held by Allen screw, or epoxied on; pieces of card between disk and hub are used to true up disk, to make rotor run true. (It may run true enough without a hub.) The neutral brush assembly has a Plexiglas block, 1/2 by 3/8 by 1-3/8, holding a brass rod 1/8 by 2-1/4 inches. The threaded screw hole at the block's other end is 7/8 inch away from the rod hole. The far end of rod is filed halfway through, with a nib soldered to it, 1/8 by 1/4 inch, perhaps 1/32 inch thick. A conducting rubber brush, 1/8 by 3/8 inches, is tied to the nib. The screw holding block seen in front view, Figure 39, is 3/8 inch down from top of minor end plate, and the same distance in from the side.

A collector brush rig, Figure 40, soft aluminum, is about 2 inches long and 1/8 inch wide. The foot is bent at right angles for 1/8 inch, with the conducting rubber brush tied to it by thread. A 1/4-inch hole in the collector lets the brush stick through to touch rods at their closest approach. The piece is bent in or out for brush adjustment. (Note: Radial Dirod Junior has brushes rigged in still simpler fashion.) The right-hand collector has the head of the brush rig mounted above center; the other has it below center. The neutral connector (seen in Plate 4) is a piece of pull-chain looped around the neutral brush-holder rods.

Plexiglas

Can be had clear, in a variety of thicknesses. Lucite (another trade name) and Plexiglas are very similar; either can be used. All of Junior's parts could be of 3/8-inch stock, or even 1/4-inch, except the disk, which should not be less than 3/8-inch. Plexiglas has some excellent properties; but it is brittle (don't drop it); it is not very rigid or stiff; it must be drilled with care. Mark outlines accurately with scratch lines; rough-cut the piece on a bandsaw; smooth edges on a belt sander or disk sander. Smoothing with file does it, but the file soon loads up; teeth must be cleared by frequent brushing. For sawing, a jigsaw may load up and cause heating which causes semi-melting; a bad cut results.

Drilling Plexiglas

Use a sharp drill, low speed, and slow feed. The constant danger is overheating, resulting in scoring. Mechanics prefer cutting fluids—kerosene or soluble oil. From a lot of experience, I recommend dry drilling, light pressure; go in 1/8 inch, back out; go further; remove chips from drill as needed.

In drilling for an accurate fit to a shaft or bar, Plexiglas can hand out a surprise: the hole may be a tiny bit small at the exit end of the hole. As the drill is about to come through, it can spring the material out and away; the material later springs back. Much of this is avoided by backing the piece with hard wood or scrap Plexiglas. For high accuracy, machinists drill the hole *undersize* and follow with a reamer.

A portable electric drill can make screw holes; hold it steady, run it in spurts to keep the speed low. For larger holes, go to the drill press. WARNING: the drill may grab, take the piece away from you, start spinning it, ruin the hole, and cut your hands. *Always clamp the piece firmly.* (More on drilling shortly.)

Threading holes

To tap a hole for machine screw threads, always use kerosene. Otherwise, the tap may fill up and do a poor job.

Flatting the disk

Unless a special high grade is ordered, Plexiglas may not be strictly flat, and thickness can vary. One side needs to be flat, as a *reference* surface. One side may be flat enough; if not, grind down high areas on a sander, using light pressure and frequent stops to avoid heating up.

Making the disk

Drill the shaft hole *first.* Unless you intend to ream, use a 3/8-inch drill. Use center punch to make an indentation. Start with a small drill to enlarge the indentation; enlarge with a larger one; then follow with the 3/8-inch drill. You will, of course, have the *reference side down.*

Next, scribe (that is, scratch or mark) the disk edge. Having drilled the shaft hole, you have lost your center. A temporary wood plug will give it back again. Use a machinist's compass, or one from a drawing set, to scribe the 2-inch radius. Scribe a smaller circle, 1-7/8 inch radius, for rod holes. Use the center punch for rod holes after marking locations. Locate, using 45 degree and 30-60 degree triangles. Drill rod holes.

Complete the disk: remove excess on band saw, then carefully grind edge to the scribed circle.

End-plate mounting

Both end plates can, if you wish, be screwed directly to the base. Screw holes in the minor (front) end plate must then be larger than the screws, permitting a bit of rocking to bring the two end plates and shaft holes into alignment. This is handwork, to overcome inaccuracies in base and Plexiglas parts. A refinement is to add the *metal* plate, Figure 39, with it screwed to the base and minor end plate permanently attached to it by machine screws. This lets the end plate be removed any number of times, with accurate replacement.

Handworking the bearings

Steel shafts normally have metal bearings—bronze, sintered bearings, ball bearings, etc. Our simple Junior has the shaft running in holes in the Plexiglas; with a drop or two of oil once in a while, these bearings last for a long time. Having drilled shaft holes in the end plates, you may be lucky and find that the shaft easily slips in each hole; or, depending on how the drill happened to behave (or if used, resharpened, and undersized) it may be a tight fit.

In either case, mount the end plates. Even with holes that slip the shaft individually, hole misalignment may be present—the shaft won't turn, when put through both holes. Handwork is called for again. Take a rod of some sort, say 1/4 inch in diameter; wrap a strip of grit cloth around part of it; put the rod through both holes; work it back and forth while slowly turning it; keep removing it and trying the shaft; likewise for the other hole. Soon, the shaft is ready to run free. Wash out the holes to remove any grit. The lives of your ancestors were full of this kind of thing: using handwork to bring an inaccurate job to near perfection. Doing a job like this will make you feel good, too!

Disk mounting

My Junior's disk fitted tight on the shaft, but it wobbled. I enlarged the hole to give enough play. Then, with the hub having a tight fit, I screwed it to the hub; it still wobbled. This called for *shimming*—placing a strip of card between disk and hub—to make the rotor run sufficiently true. (A different stunt is required when we get to Radial Dirod Junior.) The hub can be eliminated by simply epoxy-gluing the disk to the shaft, if you make sure somehow that the rotor is held in true position on the shaft.

Rounding rod ends

Using a lathe, my routine is to chuck the rod in a collet; with high speed, use file to bevel end at about 45 degrees; shift file to reduce the two ridges; then do a filed rounding to hemisphere shape; finish with grit cloth, followed by worn grit cloth or crocus cloth or fine steel wool. Lathe lacking, do likewise, using the electric drill to hold and turn the rod.

Using handwork entirely, first file the end, adopting a routine that avoids a lumpy, irregular ending, rather than a good hemisphere shape. Finish by cupping the grit cloth in the hand to fold it, stick the rod into the valley of the crease, rapidly wipe the rod back and forth while frequently turning it.

Rod mounting

Here again, your drill may give anywhere from a tight to a too-loose fit. If tight, use the drill press *as a press*, to force rods in—*but* with moderate pressure. Too tight a fit may stress the Plexiglas, cause crack-out and an accident while running. In that case, polish down the rods a bit. If too loose, fix rods into holes with epoxy. WARNING: never, later, use compounds like brass polish to clean up rods; some will creep in around rods, attack them, expand, and crack the Plexiglas.

Collector plates

These are 6-inch aluminum *blanks*; arts and crafts people hammer ash trays and such out of them. Mine happen to be 0.042 inches thick. Edges are carefully rounded by file strokes, then further smoothed and polished. Sheet aluminum this thick or thicker may be found at hardware and plumbing supply firms. Using such, make rectangular plates, 5 by 6 inches, corners rounded to 1-inch radius. If only thinner material is available (such as to cut plates from bottoms of pie pans), edge corona leakage may be too high; add *corona trim* as in Dirod II, using heavy copper wire, or 1/8-inch rod stock, applied to edge and epoxy-glued to plate.

Inductors

Aluminum or brass, ends rounded and polished. One end of the support rod should be drilled, and threaded for screw attachment to collector. Hollow out the other end to fit round inductor: file a V into it with a flat file; finish with rattail file. Epoxy-glue to inductor; use silver paint to bridge epoxy.

Terminals

Copper tubing, widely available. Shops selling it have a bending tool to bend these ears smoothly to 1/2-inch inside radius. Or use larger radius, bending around something round. Hard tubing can be flame-softened. Fill one end of an ear with epoxy, to be drilled and tapped for attachment by screw. Another option: instead of tube, use ears of 1/8-inch rod; one end to go through plate; thread for a nut on each side of plate.

Corona shields

Rear ends of disk rods come under the rear end of shields. Therefore, Plexiglas supports for shields are part of anti-corona protection: the epoxy glue holding shield to support the piece must be complete, or gap-free—a hole through it would permit spark-over.

If you learn to heat and bend Plexiglas, you can bend the shield, and make it and the support a one-piece job.

Brushes

Neutral brush assembly lets block swing about screw to adjust brush to touch rods; the rod supporting the brush can turn in its hole to make brush radial or nearly so. In Dirods I and II, and Junior, the best brush position is not directly inside of the inductor, but away from it in rotational direction by nearly a rod pitch. In the dark, you will see a brushy discharge, sometimes 1/4 inch long, between the brush and disk rods. Thus, rods begin to get charged *before* actual contact with the brush is made.

Brushes I now use, with great success, are cut from what the maker calls Slipknot Semi-Conducting Tape "used as shielding to reduce stresses around irregular points in high-voltage power cable splices and terminations." It is No. 5360 Cured Unsupported Butyl Rubber, 0.015 inch thick, made by Plymouth Rubber Company, Inc., Canton, Mass. Go to the manager's office in any electrical power utility; state your need; a couple of inches of the tape should soon be in your hands.

Until the above is secured, temporary wire brushes will do. Buy a suede shoe brush. Wreck part of the wood backing, get out a tuft of wires, which is bent into a V. Sort out about six wires. They spring apart; tie them together with thread, 1/4 inch from V bottom. Tie bundle to brush nib. Do *not* trim; leave as is, with tuft about 3/4 inch long. Shorter wires soon fatigue and break off. While writing this, I tested the wire brushes for a full two hours of running; they were still in perfect condition.

Pulleys and belts

Sewing machine agencies carry V-belts. To use such, you may need your machinist friend, to make a V-groove Plexiglas pulley.

O-rings, intended to be gaskets for vacuum work, make good Dirod pulleys. Physics departments and many research laboratories stock them, 1/8-inch diameter, in a large range of lengths, and round section. They do eventually break; keep a spare. The pulley should have a half-round groove. (I was so anxious to get my Junior into action that I first gave it a flat pulley, with a rubber band for a belt. It worked.)

Screws

Except for wood screws, all are 8-32 machine screws—all being round headed screws except for Allen screws. Allen screw: a hardened steel set-screw for holding parts on a shaft; it has no head. It screws right on down into the threaded hole. The outer end has a hex cavity, to take a hex rod that acts as a screw driver. The end-stop pieces, the pulley, and the hub all have Allen screws. Ordinary machine screws, with heads sticking out, can be used instead; but they may slip on the shaft unless you file a flat on the shaft.

Rotation

My five Dirods all run clockwise. Suppose you build your Junior Dirod, later get hold of a secondhand motor, and find that it runs the wrong way! No harm done. I have designed Junior to be easily reversed. You can easily figure out how to change the parts around for counterclockwise rotation. You will need to drill new screw holes for corona shield supports.

I have had Junior up to 1500 rpm, on test. For safety, you should run an overspeed test on any such new machine at twice the speed you ever expect to use, and take proper precautions while doing it.

Motor; speed control

All Dirods can be hand-cranked, if a proper large pulley and handle are provided. It is far better to have a speed-controlled motor. The proper motor is the "universal"—it runs on DC or AC. It is used on sewing machines and malted milk mixers. For best speed control, use Powerstat or Variac, rated at 1 ampere. Another way: adapt the resistor control of a sewing machine. Still another: rig two or three or more lamp sockets in parallel, with the entire bank in series with the motor. Experiment with bulbs of different wattages, to get a suitable range of speeds.

Corona test

Take the new Junior into a completely darkened room and run it. At first, only four little discharges at the brushes are seen. As the eyes become adapted to the dark (for some, perhaps fifteen minutes), more and more corona appears. Try everything. Draw sparks. Touch one terminal. Touch the other terminal. If it is corona-balanced, you find that coronas will shift around in odd and spectacular ways. This will be exciting, and great fun. The first thing you know, an hour has gone past. If good corona fails to develop, observe close from *all* angles. A sharpness somewhere may have been overlooked. Corona blooming there gives it away.

Current output

As noted in the Dirod Family Data Table, Junior Dirod furnishes 2 microamperes at 20 kV and 600 rpm. Current depends on both speed and voltage. Examples: at 40 kV and 600 rpm, current would be 4 microamperes; and at 20 kV and 1200 rpm, again, 4 microamperes; at 40 kV and 1200 rpm, 8 microamperes.

Dirod permanency

Let me predict. Build Junior in your youth. Put it away in a closet, go out in the world to make your way. When you retire fifty years later, come back, take it out, and run it. It will welcome you with sparks. This story is wrong in one respect: you will not put it away.

Paraffined wood for Junior

All of the foregoing is based on very extensive personal experience. Since I have not built a generator without Plexiglas, the following is suggestive. Using paraffined wood instead of Plexiglas, make a disk 3/4-inch thick; better yet, make it of two pieces, 3/8-inch thick, paraffined separately and screwed together with grains crossed. Make end plates, 1/2-inch thick. Bearings, brass plates not less than 1/8-inch thick, are to be screwed to end plates. Use glass corona shields, epoxy-glued to wood support pieces *before* paraffining them.

Building Radial Dirod Junior

Remarks

The last of the family to be born, this is the simplest and sweetest-running of them all. The Dirod Family Data Table will tell you that it equals its rival, Dirod Junior, in current, and tops it in voltage; in current, it is close behind Dirod II. The data table gives most of the specifications, and the preceding notes on Dirod Junior cover many construction features. Thus, these notes can be brief.

Attaching rods to disk

There are no rod holes to drill: rods are epoxy-glued to the face of the disk. A drawing, with radial lines for each rod position, was laid on a flat surface, and the disk laid on the drawing. Two-inch rods extend in onto the disk face by 1/2 inch. Loose rods laid down thus would tilt down to rest on the drawing; therefore, short pieces of rod were laid around to support the outer ends and make rods level. Taking up a rod, I dipped 1/2 inch of one end in epoxy and laid it on the disk—that part of the disk having been roughened to give glue better chance to hold.

Note this: at first, twenty-four rods were installed; later, the in-between spaces were rodded, making forty-eight rods, which doubled current output.

Making disk run true

Hub, 1/2 by 2 inches, Plexiglas, forced onto shaft. The shaft hole in the disk should be large enough to let the disk be tilted a little. With the disk on the shaft, the shaft was mounted in bearings. The shaft outer end was placed against something rigid to fix shaft position axially. The disk wobbled when the shaft was turned. A block of some sort was set on the table, close to disk rods. Disk tilt was adjusted until slow turning showed the rods running true, past the block. Four little dabs of epoxy then were applied, equally spaced around hub, to glue it to the disk.

Main end plate

This Plexiglas sheet is structural in function; *it also serves as corona shield.* It was intended to be of 1/8-inch thickness; our shop's stock of that was not flat enough; I used 3/16-inch sheet. The thinner sheet would bring inductors closer to rods; that should increase rod charge induction and raise current output.

The end plate was screwed to base, but this alone was structurally insufficient. Stiffeners were needed, one at each side: 1/8-inch by 1-inch Plexiglas braces, ends heated and bent at 45 degrees (see Plate 5). Screw holes were made large enough for adjustment of end plate to be parallel to disk.

Inductors

Aluminum, 3/8 by 2-1/4 inches. The inner ends are at 2-1/4-inch radius. These were epoxied to end plate. If the largest rod stock you can get or manipulate is 1/4 inch, use it. I tested such. They limit top voltage to 45 kV, reliable sparking voltage to 35 kV; but you still would have a fine machine. Later, they can be knocked off in favor of larger inductors.

Connections

Connecting wires, inductors to collectors and terminals, flexible insulated wire. Household rubber-covered extension cord will do; it commonly comes as a pair, with rubber web between. The pair can be pulled apart. Wire insertion into inductors and terminals: I drilled a short hole big enough to get wire insulation *inside* of it, went further with smaller drill for the bared wire end; then used an Allen screw to hold the wire. You can avoid the screw: *jam* the wire into its small hole (for certainty of contact) and use epoxy.

Brush arms

Plexiglas strips, 1/8 by 1 by 3-1/2 inches, held back from end plate by 3/8-inch Plexiglas blocks. They swing up or down for brush adjustment, and swing out of the way of the rotor when it needs shifting back for cleaning. These are for neutral brushes.

Collectors

Dimensions in data table footnote. Rear plates held back of end plate by 3/8-inch blocks. A screw through it connects front and rear parts of a collector.

Simplified brushes

All four brushes are at a 3-inch radius; all stick through from the rear, through 1/8-inch holes in neutral brush arms, and in real collector plates. These are double-strip brushes. Each is made of two strips of conducting rubber, 1/8-inch wide, tied together. The tie comes through the hole by about 1/8-inch. On the outside or rear surface, the two strips are *bent apart*, flat against the surface, and taped to surface. Inner or contact ends are made too long, then trimmed to rub properly. These brushes wear so very well that one adjustment lasts for many operations.

Remember: every Dirod needs a neutral connector. A machine screw is placed on the brush arm so that a nut on it screws down on a brush strip. The neutral wire, which can be *bare*, attaches to the two screws.

Screws

All machine screws are 6-32.

Plexiglas bearings last amazingly well. Serious wear could be handled later by new Plexiglas bearing blocks glued to end plates.

SAFETY GUARD. Accidents happen. Example: a neutral connector might come loose, swing in to tangle with rods, start a jam, cause rods to break loose and be thrown—perhaps into an eye. A safety guard is recommended, made of Plexiglas, celluloid, etc., as a 2-inch-wide band up both sides and across the top of the main end plate, reaching back 2 inches.

Clearance

There are two Plexiglas end-stops on the shaft, held by Allen screws; one is in front; the rear one is also the pulley. Set to give little or no end play, and to give about 1/32-inch clearance between rods and main end plate.

Further possibilities

For reverse rotation, reconnect the front of the generator.

Corona study indicates that if inner ends of inductors, and front collector plates (and other parts as corona may show up), are heavily coated with Krylon or corona dope, top voltage *might* be brought up to the 80 kV region for steady spark rate.

Another idea: shift everything that you need to have shifted, by 45 degrees, to put collectors, inductors, etc., into the *corners*: it would increase the corona shield effect of the end plate. And it *might* permit making the rods an inch longer (with other parts changed accordingly), and *might* raise current output by half. We never know how well such predictions work, until the idea is tried.)

Another idea: glass windows for charge induction. Sometime I want to try the following: cut window holes, 1-1/2 by 3 inches, through the end plate at the inductor locations; then epoxy-glue 1/8-inch glass windows against the rear, with 1/2-inch overlap. Glue inductors to windows; move rotor back to suit. There are two reasons for trying this:

First, proximity: this is the only Dirod with rods always running close to Plexiglas, which heavily electrifies. The effect is unknown; therefore, operation might improve.

Second, polarity reversal: when shorted, this Junior reverses polarity, which *may* be due to strong charges on Plexiglas at the induction area. Glass would almost certainly charge up much less. Considerable experimentation with earlier Dirods tends to support this belief.

Paraffined wood

I want to build a "no-Plexiglas" machine sometime. It would have a wood disk, with rods epoxy-glued to it *before* paraffining it; paraffined wood for minor end plate; glass for main end plate. Perhaps you will manage it before I do. If so, congratulations!

Polarity reversal

In the months after the foregoing was written and the manuscript turned in to the publisher, I put in many hours of further experimentation. I found that Radial Junior not only reverses polarity when shorted: under certain conditions it may rapidly reverse. Even so, it is simple to build, and able to perform for many experiments. Much of my testing went to trying to reduce the rate of reversal. The change finally made was to move the rotor farther back from the end plate until now the front side of a rod is a little over 1/4 inch from the end plate, and the brush systems were of course moved back accordingly. Much of this work was done to nail down the cause of polarity reversal. A good deal of evidence does now seem to say that at higher voltages, ions from neutral brush discharges collect on the end plate rear surfaces opposite the inductors, and opposite to them in polarity—thereby partly neutralizing inductor effect and reducing the induced charges on the rods.

Separately excited Junior

Radial Junior's polarity reversal can be eliminated, and its output at lower voltages much increased, by building it with its own *separate excitation* (see Chapter 11). First, let us call its generator parts assembly (end plate, inductors, collectors, brushes) *a unit*. Now suppose we build a second unit and install it about 3 inches behind the first unit. And the front unit, when installed, is *turned around to face the other way*. This makes the brush assemblies of both units accessible. The inclined braces now used in the "regular" machine disappear, for Plexiglas blocks mounted between the two end plates furnish rigidity. By the way—if we wished, we could connect the two identical units in parallel, thus doubling the output at any particular speed and voltage. But that is not our present intention; and, as you will see, the units will not be identical. Instead of that, *consider the rear unit as the exciter for the front unit*. Its terminals are connected to the front unit's inductors, and these are disconnected from the collectors. The front unit is now a separately excited generator. *My tests of my Radial Junior, separately excited from another Dirod, show that at 40 kV, its output is equal to that of the big Radial; and that at lower voltages, it would exceed it.*

But we are not done yet. The exciter (the rear unit) has no need of 48 rods: two will do! Therefore it needs no disk. It merely needs a Plexiglas bar mounted on the shaft, with a rod mounted on each of its outer ends. Of course, if you first build the "regular" Radial Junior, you can rebuild it any time later, to install the exciter unit.

These added notes emphasize the fact that all Dirods are still in the development stage, and further improvement may be possible for any of them.

Table 5. Dirod Family Data

(Dimensions in inches. Notes indicated by letters in parentheses)

	RODS PARALLEL			*RODS RADIAL*	
	Dirod Jr.	*Dirod I*	*Dirod II*	*Radial Dirod Jr.*	*Radial Dirod*
Operation and Performance					
Suggested top speed, rpm	1200	1200	1200	1200	1200
Suggested usual speed (or less)	600	600	600	600	600
Topmost kV measured	60	85+	85+	70-75	90-95
Steady spark rate kV	50	75	75-80	60-65	80-85
Approx. microamp, 600 rpm, 20 kV	2	3.2	2.3	2	8
Design Data					
Shaft size	0.375	0.5	0.5	0.25	0.5
Bearings	Plexiglas	Brass	(A)	Plexiglas	(A)
Disk diam./thickness	4×0.5	6×0.375	6×0.5	5×0.125	6×0.5
Clockwise rotation	Yes	Yes	Yes	Yes	Yes
Rods	24 Brass	36 Brass	24 Aluminum	48 Brass	30 Aluminum
Rod length and diam.	4×0.125	4×0.125	4×0.25	2×0.125	4×0.25
Rods epoxy-glued in or on	No	No	No	On disk	Into holes
Rod ends covered against corona	No	Both ends (B)	Front ends (C)	No	No
Rods extend back of, onto, into disk	(D)	0.5	0.75	0.5	0.5
Rod outer surface to disk edge	0.1	0.125	0.125	–	–
Rear ends of rods to main endplate	0.25	0.5	0.375	–	–

Table 5. Dirod Family Data (Continued)

(Dimensions in inches. Notes indicated by letters in parentheses)

	RODS PARALLEL			RODS RADIAL	
	Dirod Jr.	*Dirod I*	*Dirod II*	*Radial Dirod Jr.*	*Radial Dirod*
Corona shield	Plexiglas	Glass	Plexiglas	Plexiglas (E)	Plexiglas (F)
Corona shield is sloped 30 *degrees*	Yes	Yes	Yes	–	–
Corona shield support pieces are	Plexiglas	Fiber angle	Fiber angle	–	–
Shield size (axial dimension last)	0.125×4× 5.25	0.125×7× 6	0.125×7× 6	–	(G)
Main endplate, Plexiglas	0.5 × 4.25 ×8	0.375×8 × 9.25	0.5 ×9 × 10.5	0.1875 × 12 ×12	–
Shaft center to top of endplate	3.5	4	5.25	5	–
Minor endplate, Plexiglas	In front	In rear	In rear	In rear	–
Minor endplate	0.5 ×1.5 ×6.75	0.375×2 ×6.5	0.5×3 ×6	0.25×3 ×8	–
Collector plates, aluminum	0.042×6	0.0625×4 ×6	0.0625×4 ×5.5	(H)	(I)
Plate support pieces, Plexiglas	(J)	0.5 × 0.75 ×4	0.5 × 1.125 ×5	(K)	–
Corona trim	No	No	Yes, 0.25	No	Yes, 0.25
Inductors, aluminum	0.25 × 3	0.2×0.75 × 2.5	0.375 × 2.25	0.625 × 2.25	0.5 × 3
Neutral connector	In front	In rear	In rear	In rear	In front
Neutral brush radius position	–	–	–	3	3.25
Collector brushes placed	(L)	(M)	(M)	(N)	(O)

(A) Sintered bronze bearings, permanently oiled.

(B) Tygon tubing, 0.125 I. D., 0.25 O.D., ends plugged for 0.125 in. with epoxy, slipped onto rods; rear tubes 0.375 long; front tubes 1 in. long.

(C) Front ends covered for 0.25 in. with two heavy coats of Corona Dope (applied by holding rotor with rod ends *down*, bringing small dip cup *up* to rod ends.)

(D) Rods extend equally each way, front and rear, from disk.

(E) Main Plexiglas endplate itself is the corona shield.

(F) No corona shields; instead, inductors are encapsulated in Plexiglas tubes; tubes plugged at inner ends for 0.5 in. with RTV 102 Silicone Rubber.

(G) Shield tubes 0.5 I.D., 1 in. O.D., 6.25 long.

(H) Front plates 0.0625 × 1.5 × 2; rear plates 0.0625 × 1.5 × 2.5.

(I) Front and rear plates alike, 0.0625 × 1 × 4, with 0.25 copper tube corona trim; outer ends trimmed with 0.5 × 2 aluminum pieces; plate pair rigidly joined by brass piece of rod 0.5 × 0.75 long; front plates attached to Plexiglas blocks 0.5 thick, 1.5 axially, 2 in. long; these blocks attached to the 0.5 × 2.5 × 11 Plexiglas slant bar.

(J) Plates screwed directly to edges of main endplate.

(K) Front plates against front of endplate; rear plates spaced back by Plexiglas block 0.375 thick.

(L) 0.5 in. in front of disk.

(M) About 0.25 or so behind disk.

(N) At rear, about 0.5 out from disk edge.

(O) About 0.625 in from rod ends.

Part 2

The Dirod Manual

1. Introduction

History

In the Sixties, when looking ahead to retirement at the end of 1963, a boy-hood interest in electrostatics was rekindled, and I set out to invent and develop a new form of electrostatic generator. Since the rotor would have rods as the charge carriers, the rods to be mounted on one or more insulating discs, I coined the name Dirod, taking *Di* from *disk*, adding it to *rod*, and making Dirod. Pronounce Di—as in the word *high*. Thus, the original Dirod was born. And how did it at first perform? The poor thing would only produce about 8 kV (8000 volts). But it worked, and I was thrilled. At the time, I was almost totally ignorant of ways to eliminate spark-over internally, and to reduce limitations due to corona. By trial and error, these factors were overcome, and by gradual steps of rebuilding, my Dirod, called Dirod I, rose to a maximum voltage of 80 kV, and sometimes more. See Plate 1 on p. 35.

There followed other designs, to a total of five, two being radials instead of having parallel rods. These are described in my Doubleday book, *Electrostatics*. The book is out of print*, and unobtainable; but it will be found in a great many school libraries. Where it is available, Dirod builders should read it.

This Manual represents my experience of nearly 20 years in building and operating Dirods and demonstration accessories, and out of that experience we can say this: first, Dirods are the most rugged and reliable electrostatic generators ever produced, of the tabletop type; second, if you build a Dirod and do it right, you can know ahead of time that not only it *will* work, it *has* to work.

These retirement years have been "electrostatics years." Using my Dirods and accessories, I have travelled 145,000 miles, presenting my *Electrostatics Lecture-Demonstration* (EL-D) to thousands, throughout the U. S.; into Canada; and in three overseas trips to the British Isles. In going overseas by air, I can take only one generator, Dirod I. For such engagements, one generator *has* to be counted on to perform. No other kind of generator has such reliability. The EL-D is now almost certainly the world's best known mobile science-engineering lecture-demonstration.

---

*This book has been reprinted as Part I of the current book.

The *Manual* first deals with the materials, processes, and strategies for constructing the Dirod. That is followed by descriptions of 25 demonstration items.

No lathe work is needed in the Dirod design presented herein.

The Teaching of Electrostatics

In grade schools, high schools, colleges and universities, the electrostatics part of physics seems to be poorly taught—or even neglected—in nearly all. Teachers who try to give demonstrations and have them fail do not want to repeat such experiences. Also, many teachers who are unaware of the importance of applications of electrostatics may all the more tend to drop teaching by demonstration.

Consider three great areas of electrostatics. First, the *separation of mixtures*: in the mineral industry, about thirty kinds of ores are separated and refined electrostatically, to the extent of millions of tons per year; otherwise, other applications are at work, such as cleaning up mustard seeds to make them fit to eat. Second, *electrostatic precipitation* traps fly ash made by coal-burning powerhouses to the extent of 20 million tons per year in the U. S. alone, and operates to keep other pollutants from pouring into the air. Third, in *electrostatic coating*, spray painting saves some 50 million dollars per year in reducing loss of paint; it applies powder coatings, later to be fused onto the target—such as your refrigerator and other household and yard-use devices. Every bicycle in the world is coated electrostatically, including mine. *Xerography* is an electrostatic process. A measure of its success is seen in the business done by the Xerox firm alone, now around seven billion dollars per year. For more about such matters, see my book mentioned above. Also, see my article, 'Electrostatics' in *Scientific American* for March, 1972.

For teacher and student, electrostatics is fun. I claim there is more fun to be had in electrostatics than in any other area of science!

Feedback

It is my hope that after teachers have adopted Dirods and demonstrations, and students have entered Science Fairs with their electrostatics exhibits, I will be hearing from both groups. One of my main pleasures is to get this kind of feedback. My correspondence is very heavy for a number of reasons, and I regret that I may not be able to reply to all such letters.

Science Fairs

Can students build their own Dirods, enter them in Science Fairs, and win ribbons? Not only they can, *they have*. Over the years, I have helped students, to the extent of a hundred or more, do just that. And one way to have a Dirod do double duty is for a teacher, needing a school Dirod, to encourage a student to make it, enter it in a Science Fair, and then hand it back to the school. The school need only pay for the materials. The student, of course, in his exhibit, will acknowledge the aid furnished by the school, and such advice as may be given in the project by the teachers, or others.

The student entering a local, regional or national Science Fair with a good electrostatics exhibit is due for a very interesting experience. Your demonstrations will be so interesting, and so much fun, that other exhibitors will be coming to your booth to share in it. And there is plenty of time for that. I know. I have been a judge in a U. S. National; and twice, a judge in the Canada-Wide Science Fairs. There is lots of waiting at times, for the judges to get around. Now, no judge—a teacher, or engineer, or scientist—knows everything. And hear this: your chance of having one of them being expert in electrostatics is slim. The chances are that the judges will take an unusual interest in your demonstrations, and in the quality of workmanship shown in your apparatus. There is also a risk, and it is this: they may give you more credit than you deserve.

To explain this statement, remember: in science, "We stand on the shoulders of those who have gone before." This means that in your displays you must give credit where credit is due. If a teacher or a mechanic has been of aid in your project, say so. If your school contributed materials, say so. If your Dirod and demonstrations grew from using this Manual, say so. Don't run the risk of having any judge jump to the conclusion that you invented Dirods. You didn't. I did. In all this, you lose nothing. Instead, you gain respect by showing that as a young scientist, you know how to be fair "... to those who have gone before." If you build your Dirod from this Manual, the Manual should be in your exhibit.

Safety

Your Dirod will be completely safe, and the same goes for when it is used with any of the demonstrations described herein, no matter how or how often you expose yourself accidentally or on purpose to the electrical discharges. If a man wants to commit suicide, he can climb a tower and reach for a conductor in a 50 kV (50,000 volt) power line. He doesn't need to shake hands with it. It will accommodate him with a killing arc. In contrast, if your Dirod is humming along, unloaded, and making 80 kV, and you approach the terminals with the knuckles of both hands, all you get is a stream of little sparks, causing a tingle. (A warning about *ozone* will come later.)

You *can* work up a way to be unsafe. That is, to build a large capacitor, charge it up to maybe 50 kV, and let someone be accidentally exposed to a killing discharge of energy. This, of course, can be done with the generators on the market, and now present in thousands of schools. Any of these, including Dirods, could charge a large capacitor, and deliver a killing discharge.

2. How a Dirod Works

How a Dirod Charges Up

The schematic drawing of an elementary Dirod, Figure 1, is taken from my book, *Electrostatics.** Only the electrical parts are shown. The collector

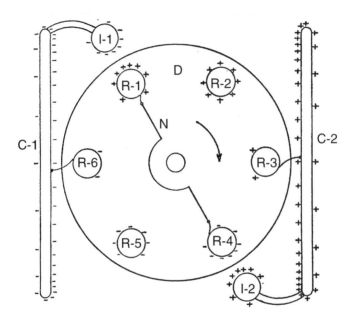

Figure 1. Dirod schematic drawing

plate C-1 is connected to the inductor I-1. Collector C-2 is connected to inductor I-2. The rods R-1 and R-4 are connected momentarily by way of the neutral N and the brushes on the ends of N. The negative and positive charges on I-1 and I-2 now have an influence on the two rods, charging them *oppositely* to the inductor charges. I use the word *influence* deliberately, for in earlier times, it was used, and any such machine was called an influence generator. Then the word *induction* came into use, and we now say, induction-type generator. We say that negative C-1 has *induced* positive charges on R-1.

---

*Part I of the current book.

As the disk rotates, the rods carry their charges to the brushes on the collectors and thereby add charges to the collectors. Thus the collector voltage builds up and up, until some limit is reached.

But now, you will ask, how does all this get *started*? When you build your Dirod, using Plexiglas for insulating parts, you will handle it in various ways. Drilling it, cutting it on the bandsaw, sanding the edges on a belt sander, handling it—any of these will cause some charges to appear, somewhere on the Plexiglas. These will be irregular, and unknown as to strength and location. But they will almost never cancel out their effects. Some charge will prevail to give a starting-up effect. In short: build your Dirod. Finish it. Turn it on for the first time. A thousand to one, it will generate! I have started up my Dirod I thousands of times. Only three or four times has it failed to generate, probably when it had gotten too dirty, and high humidity had let most charge drain away. Letting the rods bat against a limp plastic bag then adds some frictionally-developed charges to act as starters.

Buildup is Geometric

There is a vast difference between the way an arithmetic series of numbers, such as 1, 2, 3, 4, ... builds up, as compared with a geometric series such as 1, 2, 4, 8, ... Suppose a Dirod has built up to one kV, and that in one revolution it goes to two kV, then three, etc. Slow buildup. But it works faster. If, in one revolution, it goes from one to two kV, the inductors, having doubled their voltage, will double the rod charges, the next revolution will double again, and the rise in generated voltages is like the numbers 1, 2, 4, 8, 16, 32, 64... Plainly, this is a runaway rate of rise. It means that if you short a Dirod and drop its voltage down near zero, the time it takes to achieve full voltage again is amazingly short.

Electrical Discharges: Sparks and Corona

Air breaks down at 30,000 volts per cm. Consider the sphere gap, used for measuring high voltage. As two metal spheres at, say, 20 kV are made to approach, the electric field between them rises. It is highest where they are most nearly together. When the field intensity reaches the above limit, a *spark* passes between them.

Now think of an aluminum plate 1/8 inch thick, 6 inches long, 1-1/2 inches wide, and rectangular. Isolate it, and bring it up to 20 kV between it and the surrounding room walls. The field intensity at its surface will be lowest at the center of a flat side. As you move toward an edge halfway along the length, the intensity rises rapidly; or toward an edge at the end, still more rapidly. At the sharp edge, it will be very high. Next, move toward a corner, and the intensity at what amounts to two points will be still higher. There being no conductor nearby to spark to, there can be no spark. But do this all in the dark, and you will see *corona* at and near these sharpnesses. The air breakdown strength has been reached, the air molecules are ionized, and a corona current occurs. Now remove the plate, round the corners to a half-inch radius, and try again. Field intensity at corners has been reduced, but the sharp edges still are corona sites. Next, get rid of those edge sharpnesses by rounding the edge of the plate all around. The field intensity will still be highest at the four rounded corners, but at 20 kV, may not be high enough to reach 30,000 volts per cm, and no corona may be visible. The plate just described is one of the collector plates of your Dirod.

Observe Figure 1, and note the way charges, plus and minus, concentrate on corners and edges of the live parts, as compared with how they become fewer on wide flat areas. The lines of electric force in the electric fields begin or end on these charges. The higher the lines of force are concentrated, the higher becomes the field intensity, and the tendency of the air to break down, tending to spark over, or cause corona to form.

Ions

Ions are charged atoms or molecules. The normal (or uncharged) molecule has as many electrons (minus, or negative) as protons (plus, or positive), and is neutral. If a molecule loses an electron it becomes a positive ion. If it gains an electron, it is a negative ion. In the atmosphere, cosmic rays from the universe and radioactivity of minerals in the earth are constantly producing both kinds of air ions. A cubic centimeter of air contains billions and billions of molecules. Only a few natural air ions are produced, there being only a total of a few thousand in pure country air.

In an electric discharge from a negative corona point, or from any surface where the field intensity is great enough, great numbers of negative ions are produced, and go streaming away. Likewise for positive ions produced by positive corona points or overstressed surfaces.

If you set up two electrodes a few inches apart, rod-shaped, with rounded ends, and apply say, 80 kV, the corona glowing at and near each electrode betrays the existence of streams of ions. At the negative electrode, great numbers of negative air molecules are being produced and, acted on by the Coulomb force, are flowing in the electric field to the positive electrode. These charged molecules carry neutral molecules with them, and thus the electric wind is produced. At the same time, positive air ions made at the positive electrode are streaming toward the negative electrode, also making the *electric wind* there. The two streams of ions constitute a *current*.

Watching Corona

When your Dirod is finished, you will operate it in a totally dark room at its maximum voltage, 80 kV or higher, and watch corona. Various charged parts are now close together; the electric fields are extremely complex; and you will see streams of glowing corona here and there. All of the current generated is then being used up within the machine, by way of corona discharges. *Warning: ozone is certainly being produced, and this, at sufficient concentration, is a poison.* Taken all together, I have watched this corona, enthralled, for many hours in the dark, in a small, unventilated space. Apparently, I suffered no harm. But I do recommend that you keep these episodes brief, or ventilate, or both.

Designing for High Voltage

Suppose you were designing a simple nonelectrical machine, such as a fan. It would be an ordinary mechanical engineering job. Provide a shaft, bearings, a hub on the shaft to hold the vanes; add a pulley for the drive, and that's it. But to design a high voltage generator is a totally different ball game.

Suppose you took the schematic drawing, Figure 1, and built it just as shown. It would generate. But how well? It might make all of 10 kV, at which voltage it would limit itself by sparking between inductors and the passing rods. So then, you cure that by interposing spark shields of glass or Plexiglas and kill the spark-over. You might then get 20 kV. Why not more? Take it into the dark room, and you see it has gone into corona. The spark shields, just big enough to stop sparking, do not extend far enough each way, and corona now limits you. You then extend the shields, as better corona shields, and perhaps 40 or 50 kV is the outcome. With so few rods, the current output would be very low, but at least, it does generate.

The design of a successful Dirod as a compact machine of moderate size, able to deliver plenty of voltage and current for numerous demonstrations, meant placing live parts close together, but not too close; and being hampered by the fact that the electric fields are extremely complex. This means that there cannot be a way to carry out the design, step by step, as in cases of simple mechanical design. Instead, slow advances had to be made, *experimentally,* for me to make all of the changes needed to bring Dirod I up from its pitiful beginnings at around 8 kV maximum, to 80 kV. It was, and is, "the original," the pioneer Dirod. I am very fond of it. It has been my unfailing companion on every one of the 145,000 miles of my retirement electrostatics career.

When I had brought Dirod to the stage shown, Plate 1 on p. 35, it had complex structural parts, no longer needed. These were put on to make both sets of brushes adjustable. Experience proved that the brushes can be fixed in position. That much simplified the construction. The various changes over the years account for the fact that the panel of Dirod I has 32 holes through it, all now epoxy-plugged except the few needed.

Dirod Polarity Reversal

Operate a self-excited Dirod at no-load (nothing connected to its terminals) in the dark. You will see a discharge glow at the neutral brushes. Why? The generator has gone up to maximum voltage, maybe 80 kV. Half of that, 40 kV, is between an inductor and the rod making contact with a neutral brush. Contact? At such a high voltage, the current charging the rod doesn't wait for contact. The very intense field causes most, maybe nearly all of the current to go by the discharge path in air. That path is rich in ions, probably of both kinds, plus and minus. Looking at Figure 2, suppose the upper inductor is negative, the rod positive. Any positive ions will tend to move toward the inductor. But the corona shield is in the way. Therefore, it promptly is plated, in the immediate area, with a positive charge. This is what I have called the "adverse charge." It is unwanted for two reasons. First, it *reduces* the rate at which charge is put on the rods. Second, if we short the machine (say, by touching knuckles to the terminals) the collector and inductor charges disappear, and the adverse charge takes over and reverses the machine's polarity. It promptly builds up again. This disadvantage is inherent. However, for many demonstrations, it is only a minor effect one soon allows for.

As we shall see, polarity reversal can be eliminated by making the Dirod, not self-excited, but *separately excited*.

3. Materials, Sources, Methods

Brass and Aluminum

In theory, many kinds of metals and alloys would serve. In fact, only two need consideration: soft aluminum, and brass of the softer grade rather than hard brass. These are the *only* softer materials that are made in a wide variety of thicknesses and diameters. If there is a general machine shop in town, go there. You might be lucky. If there is a hobby shop in town, start with it, and be directed to one of the thousands of hobbyists who have home workshops (and make friends while there: you may be allowed to do your work there). The hobbyist may not have all the materials you need, but may know how to get them. In a large city, look in the Yellow Pages of the phone book for words such as Copper, Brass, etc. The city has one or more outlets for basic materials, where your needs may be met, out of stock. If they do not have brass for rods and must place an order for you, you may run into a high minimum charge. In that case, walk into a hardware store and buy Bake Rods. These are the soft aluminum rods used in your kitchen, to poke into big potatoes and such to carry heat into the interior. They are larger than the 1/8-inch rods described herein, so, use 30 instead of 36. Enough will cost under $10.00. If you cannot locate flat aluminum for inductors and collectors, I will tell how to make acceptable substitutes later.

Inductors and Collectors

Collectors are made of 1/4-inch soft aluminum. If unavailable, use two pieces of 1/8-inch, epoxied together, and, epoxied before being rounded and smoothed. Mark the outline, rough out the piece on the bandsaw. Corners are to be rounded to 1/2-inch radius. Next, use the belt sander, carefully rounding the corners and edges. This gives a rough finish. Clamp the piece in a vise. See-saw down over an edge with emery cloth for a better finish. This will reveal high places. Take these down with strokes of a file. Follow with more work with emery cloth to get a very smooth finish. Then make it smoother still with a piece of crocus cloth, or steel wool. Inductors are similar, but made of 1/8-inch aluminum.

Rotor Rods

Brass or aluminum. The ends must be given smoothly rounded ends, approximately hemispherical. This can be speeded up. After cutting to the 4-inch length, chuck a 1/8-inch rod (for a 36-rod rotor) into an electric drill. Clamp a piece of Plexiglas in a vise, with a hole through it. Stick the rod end through the hole. Start the drill. As the rod spins, use cross strokes of a file to round the end. Get a smooth finish with emery cloth. With 1/4-inch rods (for a 24-rod rotor) a first rough rounding can be done on the belt sander, prior to finishing with the electric drill procedure. (Using Bake Rods, make a 30-rod rotor.)

Brush Material

Furnished herewith.* Use sparingly. Conserve it, treasure it. It is butyl rubber, made somewhat conductive. It wears, and wears, and wears. Make the brushes 1/8 inch wide, or a bit wider. Do not cut with scissors; a ragged cut would result. Lay the piece on a hard surface. Lay a piece of smooth-edged metal on it, press down hard, and cut along the edge with a sharp knife-point. Make the strip somewhat longer than needed; trim back after installation. The remaining rubber: tape it to the base of the Dirod somewhere, to make it permanently available.

Plexiglas

Plexiglas is essential for the insulative parts of a Dirod, and for structural parts of demonstration items. Fortunately, it is available, countrywide, in builder supply stores. Widely used for wintertime storm windows. A square foot of the 1/8-inch thickness may cost about $1.50. Some stores will carry it in 2 by 4-foot sheets; others may have other sized sheets also. Some may also carry it in 1/4-inch thickness. I have found one store having only a thinner sheet: 1/10 inch thick.

This thinner kind is not nearly stiff enough for the Dirod bearing posts, and the panel, but here, it can be doubled to 0.2 inch by laminating two sheets together with epoxy. The two rotor discs need not be thus doubled: they can be each 0.1 inch, for when the rods are epoxied in, the rotor will have plenty of rigidity.

If your store carries Plexiglas of 1/8-inch thickness, and also 1/10-inch, you could use the thinner material for many of the structures for the demonstration items. It is more easily worked up, as in bending, for example.

Cutting and Dressing Pieces

Plexiglas has its own special properties. Honor them and you win. Fail to do so, you can lose. In fact, when a skilled machinist first works with it, he can have trouble.

---

*AD's original supply is gone, but equivalent material can be obtained in some ESD protective wrist straps, available at electronics supply stores.

Mark the boundaries by scoring the outline with a sharp needle. Use a bandsaw. If you never have used one, get instruction. Fingers are precious. Practice on scraps to get the feel of it. Soon, you can cut very close to the line. The rough edges are then smoothed on a belt sander.

Drilling Plexiglas

Do it on the drill press. Indent with a prick-punch to guide the tip of the drill. Suppose you have a new, very sharp drill. You start drilling. All goes fine until the drill is nearly all the way through, and then, disaster strikes. The drill *grabs*. When that happens, you cannot hold the piece down. It climbs the drill, whirls with it, and maybe bangs your fingers. The hole is now irregular, and the piece may even be cracked next to the hole. Apparently, what happens is that just as the drill starts through, it grabs onto the floor of the hole and lifts. The one sure way to avoid this is to clamp the piece down. The experienced machinist knows this; and first, he slightly *dulls* the drill with a fine-grained carborundum stone. *Do not drill continuously.* Drill down a little, back off; a little more, back off. Do this maybe four or five times, to get through a 1/8-inch piece. This avoids overheating. Plexiglas is a poor heat conductor. If overheated, the hole will be rough and irregular, not smooth. Also, *support* the piece on a smooth block of wood. Plexiglas is not very stiff. As the drill is close to going through, the piece can yield by being pushed outward near the bottom. When this happens, you have a hole smaller at the bottom than the intended hole diameter.

Again, this matter of the piece climbing the drill and ruining the hole: if you cannot clamp the piece down to the drill table, get someone who can *strongly* hold the piece down for you. Also, a ruined hole does not mean discarding the piece. You can put a Plexiglas pad on it, with epoxy, and drill it. This especially refers to the two rotor discs.

Sizes and Fits

Your Dirod's quarter-inch steel shaft will have bearings at each end, and these will be Plexiglas bearings. It is fortunate that such bearings last through almost any amount of use. So now, use a 1/4-inch drill and make a bearing hole, for practice, through 1/4-inch Plexiglas. If this were an ideal world, the shaft would now exactly fit the hole. It won't, except rarely. In fact, if the drill is worn, the hole may be undersize, and you have a force fit, not an easy sliding fit. What to do? You enlarge the hole, and do it just as I have done. Find a rod or something (even wood) smaller than the hole. Machinists all have inch-wide abrasive cloths, called emery cloth, ranging from coarse to fine. They will be glad to give you a few inches of the finer stuff. Starting at one end of the rod, wrap it spirally onto the rod for two or three inches, not overlapping. The rod must be small enough to let you insert the combination into the hole. Move it *straight* back and forth, stopping often to turn it to get uniformity. Soon your shaft will have a running fit. Flush out any grit with lighter fluid, or soap and water. These bearings need only a drop or two of oil once in a while, to give fine service.

Bending Plexiglas

With 1/8-inch stock, right angle bends are easy to make. Angle pieces are needed to hold the corona shields in place. I have worked up my own method and used it many times. Practice on scrap, of course. Light a candle. Pass the part, at the bend-to-be, back and forth above the flame, briefly. Turn over, repeat. Turn back, repeat, and so on. Soon, it is ready to bend. Bend it by hand, to get practice, and get the "feel" of the operation. Next, if you want an angle each side of which is an inch wide, cut a piece of scrap two inches wide and three inches long. Get a square block of something. An inch back of the edge, tape a stop-piece. Heat your sample. When limber enough, lay it to the stop-piece. Hold down on the half that rests on the block with something. With a piece of wood or metal, bend the projecting half down against the block's face. Hold it there until stiffness returns. Look at your sample. The chances are, you do not have quite a right angle bend. Simply reheat a bit at the bend, and bend it now slightly *beyond* what is enough, and watch it relax back. With such manipulations, you will soon learn to get the bend just right. You may find Plexiglas angles to be invaluable in constructing the demonstration items.

The angles holding the corona shields must be made quite true right angles, else the gap between shield and rods will not be uniform.

Corona Shields

These can be either Plexiglas or glass, 1/8-inch thick. (The 1/10-inch Plexiglas, when obtainable, would also do). Plexiglas is unbreakable: an advantage. But my experience indicates that glass collects less adverse charge, and thus tends to give a somewhat higher current output, and is less subject to polarity reversal. Either will serve.

Epoxy Adhesive

Epoxy is the amateur's best friend. Likewise for the professional. In the last few years a great advance was made when Five-Minute epoxy was brought out. Long, vexing waits are thus avoided. It used to be that if, say, a small piece of Plexiglas was epoxied to a larger one and simply set aside, the liquid epoxy at an edge of the smaller one would exert capillary attraction, and pull it out of position. This called for taping to make it behave. But now, with rapid setting, one can hold the piece to position as setting advances, and make it right the first time.

In joining parts with epoxy (both Plexiglas, or both aluminum, or Plexiglas to aluminum) if the parts are smooth, a very weak joint can occur, and it may fail. Roughen the surfaces. On the other hand, if a mistake has been made, the pieces can still be split apart and cleaned up and used again. Place the lower piece against a heavy block or a stop-piece. Place the edge of a chisel to the edge of the joint and give it a sharp rap with a hammer. *Wear safety glasses!*

Special note. The epoxied joint between collector and its Plexiglas support, Figure 3, should have extra strength. After making the joint, drill through both parts, making the two holes indicated, and fill them with epoxy.

Corona Dope and TV Tubekoat

Corona Dope

A thick and reddish liquid that hardens slowly. It is made by GC Electronics Co., Rockford, Illinois, and comes in a two-ounce bottle. TV repair shops should carry it. Where corona tends to develop, it can suppress the corona. The electric field at rod ends can be very intense, and unless corona there is suppressed, it gives a rich display in the dark at high voltage. Both ends of the rods are to be coated (unless another means is used). The easy time to do this is *before* the rods are mounted in the rotor. Tape a batch of rods, spaced, vertically, to the edge of a worktable. Rig a small cup with a handle, pour a quarter inch of dope into it. Bring the cup up under each rod. Allow the dope to become firm (it may take some hours) and give it a second coat... An alternative method: use Tygon tubing. For 1/8-inch rods, use tubing of 1/8-inch I.D., 1/4-inch O.D., in pieces 3/8 inch long or a trifle longer. Epoxy the tube to the rod, and slip it on for a quarter of an inch. Later, if the outer end is not epoxy-filled, fill it.

Tubekoat

Made by the same firm as above. A dead black liquid. When dry it is conductive. Same size bottle as above. The brush supplied with it is much too stiff for getting a smooth coat. Use a camel hair brush. It will be used to coat ping-pong balls conductively, for certain demonstrations.

Nonmetallic Inductors and Collectors

Coated Plexiglas

If unable to get the aluminum for these parts, there is an alternative. Make them of Plexiglas, corners and edges rounded and smoothed as already described; then coat them *all over*, conductively, with Tubekoat. This coating will not be as good a conductor as if a metal coating were put on by electroplating; but as for serving its purpose in the generator, it behaves the same.

4. Building Your Dirod

The Drawings

The drawings show an approximate scale, but not the exact dimensions of each part. To find the size of some part, *measure the drawing*. For example, the panel is 10 inches high. If your rule is used on the drawing, it may find it to be a trifle different, over or under. This merely means that in reproducing the drawing, the process has slightly altered the size of the paper sheet. Figure 2 through Figure 5 can be matched, side by side, to match the front view with side views.

Two Ways to Make a Dirod

The machinist's way

In my department at the University, we have a splendid machine shop (I often use it), and we have Jim and Otto, both splendid machinists. They know every trick of the trade. Suppose I could hand the drawings to Otto, and have him make six Dirods. He would use precision methods, and make six of each part, all alike. In final assembly, he could pick up any part and use it in any of the Dirods. All machines would fit together with high accuracy, with no adjustments needed.

Your way

Totally different. You may never have built anything this complicated. You learn as you go along. Most of your screw holes are well located, but some are a little off. Also, wood, for the base is not an ideally workable material. In making a hole for a wood screw, it can be a bit off; that tilts a bearing post sidewise, and makes the shaft out of alignment. That's when you somewhat enlarge the hole in the Plexiglas, make the adjustment, and go ahead to the next job. When done, your generator works just as well as one Otto would make.

The Base

The base, wood, an inch thick, and the sub-base (can be plywood) are cut out on the bandsaw. The front and rear edges of these, where wood screws

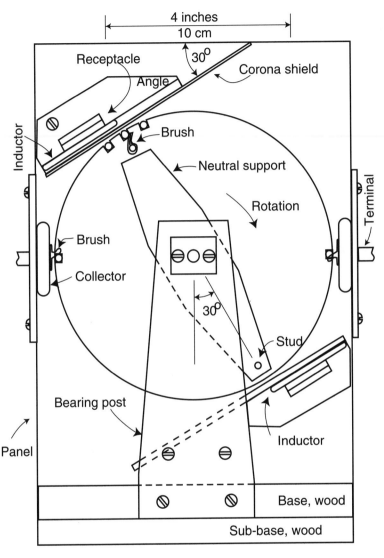

Figure 2. Dirod drawing A

hold vertical parts, should be smoothed on the belt sander, using a guide to make these surfaces pretty accurately flat and vertical.

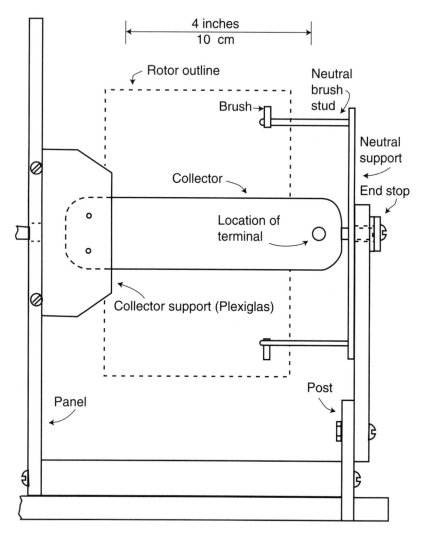

Figure 3. Dirod drawing B

Bearing Posts

Use quarter-inch Plexiglas. If unavailable, epoxy two pieces of 1/8-inch material together. The front post is a two-piece affair, and for good reason. In your building, you may need to mount the rotor, and dismount it, several times. Using a one-piece post here would call for unscrewing the wood

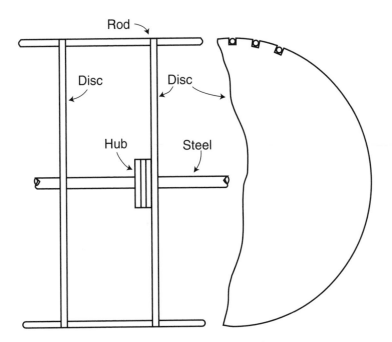

Figure 4. Dirod drawing C

screws each time—a nuisance. The machine screws holding the upper part to the lower are easily removed and put back.

Shaft, Bearings, and End Stops

The shaft should go clear through the post's Plexiglas bearing hole and somewhat beyond. The inner part of the end stop is Plexiglas; outer part, brass. The brass acts as a thrust bearing. The shaft needs a bit of end play. Therefore, *do not cut* your quarter-inch shaft to length *until* you have assembled the posts to the base, and marked the shaft. Remember: wood is not an accurate material. Your posts may lean a bit toward or away from each other. Saw the shaft long enough to allow finishing the ends smooth. With care, grind the ends close to flat on the belt sander; or, file them. Final finish to smoothness by using the electric drill and guide hole method described for rounding rod ends. Try to make the ends slightly convex, for bearing against the brass. (If brass is not at first available, make the outer piece of Plexiglas for temporary use; but later replace it with a brass plate.)

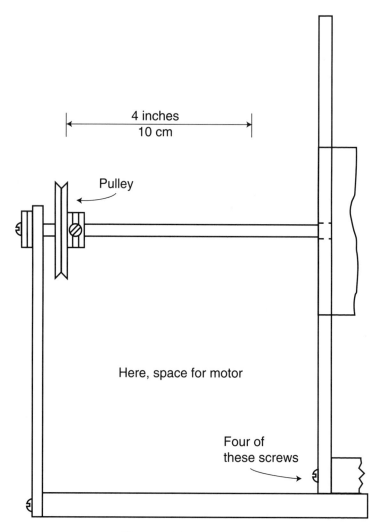

Figure 5. Dirod drawing D

The Panel

Preferably, for rigidity, the Plexiglas panel would be of quarter-inch material. But, if not available, or if too hard on the pocketbook, use 1/8-inch thickness. This is not thick enough to take the machine screws holding the Plexiglas collector support. Therefore, pad it out with small pieces of Plexiglas, epoxied to the panel. Where the shaft goes through the panel, make the hole 3/8 inch. If for any reason the 1/8-inch panel is too flexible, you can always add a stiffening Plexiglas strut from near its top, to the rear post.

The Rotor

In my Dirod I (Plate 1 on p. 35) and Dirod II (Plate 2 on p. 41), my rotors are overhung.That is, the bearings are: one in the rear post; the other in the panel. This design calls for good bearings, especially with so much overhanging rotor weight bearing down on the panel bearing. My rotors have single, thick Plexiglas discs, and with the rods inserted through drilled holes out near the periphery. The present design, herein, is very different indeed, it having two thin discs, with rods held in slots cut into the disk edges. *This is part of the design by which you can build a Dirod without using a lathe.* The notched disk can be used for 1/8-inch rods, or 1/4-inch; or for the in-between size of Bake Rods, if used. The notched rods are superior to those in drilled holes, for they are closer to the inductors, and therefore should have higher induced charges.

The rotor must be demountable. Charged parts inevitably collect smoke, dust etc., and such insulative surfaces become somewhat conductive, especially in warm and humid weather. Taking the rotor out allows it to be cleaned with soap and water, then thoroughly rinsed. The panel and the corona shields also will need cleaning from time to time.

Making the Disks

Shaft holes

Drill these first. Otherwise, if delayed, a climb-the-drill accident would ruin a lot of work. As to fit, have an easy fit, to let the disk slide easily on the shaft. But in doing so, we have "lost the center," for next comes accurately scribing the disk outline. Insert a piece of 1/4-inch rod into the hole. Take a piece of wire, make a loop at one end big enough to slip over the rod. At the other end, make a small U-bend, such that, using a needle as a scribe, the needle will accurately mark the circumference, with a radius of 3 inches.

Cutting and notching

Use the bandsaw to rough-cut, taking off material to within a quarter inch or less of the scribed line. Then much more carefully, make cuts to leave a radial excess of 1/8 inch or less, but leave some excess. Then careful work on the belt sander will bring the disk down to size. Next, using a protractor or other means, quite accurately make scribe marks radially, for one side only of the slots, as long as the rod diameter. Example: for 36 rods, these marks are 10 degrees apart. Then one by one, mark the other sides of the slots. Then back to the bandsaw, but *not* with one of the two discs. Practice first, by notching a scrap piece after marking it. Tape stop pieces at both sides of the saw blade, and adjust them so that when you advance the piece, it will cut in just deep enough to give a 1/8-inch depth for, say a 1/8-inch rod. Keep testing with a rod, to learn to cut to give it a close fit. After enough practice, cut the slots.

Mounting Discs on Shaft

Front disk

The front face of the disk will be 2-3/4 inches from the front end of the shaft. Mark the shaft there, lightly, with a file. Back of the mark, use a file to lightly roughen the shaft to make sure the epoxy used for disk and hub will stick to the shaft. Get a *flat* board, preferably not less than 3/4 inch thick, and drill a 1/4-inch hole through it centrally. Push the shaft's front end down through the hole. Hope for a fit such that it takes some force to get the shaft in, and that the shaft is *truly at right angles* to the board. Test for right angularity with a right triangle at several positions. Push the shaft down until the file mark comes just to the board. (Use the drill press for the hole; you cannot hold an electric drill with accuracy. If the hole is not quite true, try another hole, or even another, until things come out right.) Unless this is done quite well, the disk will wobble. It would signal bad workmanship. Get it right!

Now for the hub

Make the three hub Plexiglas pieces round, or square, as you like. The disk surface where the hub joins should already have been roughened; likewise for the hub piece surfaces. Apply epoxy lightly to the under side of the first piece, press down lightly, and turn it to make the epoxy uniform. For this piece, avoid being generous: we do not want epoxy to creep down along the shaft in excess enough to glue it to the board. Let set. Apply the next two pieces, one by one. With these you can be more generous.

It will be 1-7/8 inches between the inner surfaces of the two discs. Using a wood stick, cut three pieces to this length. They will be temporary spacers. The ends should be square enough so that they will stand erect, equally spaced, on the lower disk. The exact location of the rear disk is of no importance. What is important is that the height of these posts shall be alike, to make sure the rear disk has no wobble. Now bring the rear disk down onto the shaft and onto the posts. Put a rod into position. Sight past it at the shaft. Turn the disk until there is parallelism. Try the same at other slots. If all goes well, apply epoxy liberally to the disk and shaft, above and below. Before it sets, check again, to make sure the rods will not be skewed, but parallel to the shaft. (If a friendly machinist offers to turn a couple of nice brass hubs for these discs, thank him but don't take the offer. Such hubs are prone to invite streams of corona across the disk surface, between rods of opposite polarity.)

Rodding the Rotor

If you are using Corona Dope to protect the rod ends against corona, you have already applied the coating to the rods. They are ready to be mounted in their slots. Have the shaft horizontal. Put three or four rods in top slots. Devise some way to make sure the ends are very closely all in the same plane. Use the eye end of a needle to pick up a dab of epoxy and apply it to the slot and rod at both discs. Turn the rod to get the epoxy down in and around. This is a tacking operation. Turn the rotor, bring the next batch on top, and repeat. With practice gained, and using Five-Minute epoxy, you may be able to affix six or more rods at a time, as one batch. With tacking finished, go through the whole thing again, making sure that the rods are strongly epoxied in. Except for the pulley, your rotor is now finished.

The Pulley and Belt

The pulley is easily made of two 1/8-inch Plexiglas discs. After drilling them for a slip-fit on the shaft and making them matched circular discs, carefully bevel them (Figure 5) on the belt sander. The pulley must be demountable. Add a hub, as shown. Drill and tap it for a 6-32 steel screw, with the end ground to a point. Harden the point by heating to a red heat and quenching in water. O-rings make good belts. Or, hunt around for a strong rubber band of the right length.

The Motor and Speed Control

Use a sewing machine motor. If secondhand, be *sure* to test it before buying it. As far as I know, all these motors run the same way. With the pulley end at the rear, it runs clockwise. A Variac or Powerstat rated for one ampere and up to 115 volts gives perfect speed control. These are costly and heavy. For my lecture trips by plane, I use a water rheostat I have designed. It costs little, and weighs but four ounces. However, because of safety requirements, a student wishing to build it must get a science teacher to request it, and to guarantee supervision prior to its use.

The Brushes

Collector brushes each require two strips of the conducting rubber. These are bent L-shape, placed back to back, and with the feet taped centrally to the collector plate. The neutral brushes are one-piece, Figure 2. Wrap the strip around the stud, and bind with crisscross wrappings of thread. The brush can then be turned to point out radially.

The Neutral Connector

If there is no connection between the neutral brushes, no charge can be induced on the rods. The connector carries such a small current that it could be a very small wire. A neat neutral can be of 1/8-inch brass rod with U-shaped ends, these being slipped over the brush studs. It must be bent in the middle to avoid rubbing on the shaft. It can be taped to the inner side of the neutral support, Figure 2. No. 8 bare copper wire could also be used.

The Dirod Terminals

The collector plates must have Dirod terminals onto which connectors going to a demonstration can be hooked. One is seen on Dirod I, Plate 1 on p. 35. If you have 1/4-inch rod stock, use it. Bend to as short a radius as you can (see the connector, Figure 6) and mount it with the bend upright. The upright should be out about an inch from the collector. Run the machine in the dark

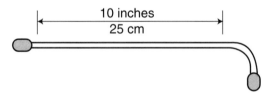

Figure 6. Rigid connector

to check for any excessive corona from the smoothly rounded top. Corona may need suppressing by use of Corona Dope, or by epoxying a Plexiglas cap down over the top. Bake Rod material could be used. It being smaller, all the more may it need corona suppressing.

Lengthening the Machine

The design is easily modified by "stretching" it. For example, with 8-inch rods, and all corresponding parts lengthened accordingly, the current output would be doubled, at the same voltage and speed. The distance between rotor discs should be about doubled. A shaft of 3/8-inch steel should be adequate, for the doubled length.

The Self-excited Dirod

The collector and inductor at the left are to be connected. Likewise for those at the right. The connectors can be 1/8-inch brass rods, with ends rounded and smoothed. A connector is given a 30-degree bend, located so that one end comes about to the middle of the inductor; the other to about the middle of the collector. Tape them into place. These go on the outsides, not inside. (Later, if you add separate excitation, these will be replaced. Also, *until* and *if* you provide separate excitation, omit the two-piece receptacles, Figure 2, attached to the inductors.)

The characteristic

A "characteristic" is a curve showing the performance of something. Curve a, Figure 7 is the characteristic of the self-excited Dirod when run at around 600 rpm. Any such curve is only approximate, for the behavior will de-

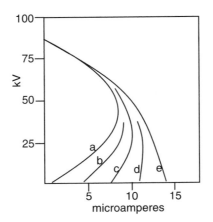

Figure 7. Characteristic of a self-excited Dirod

pend on variations in the design of the generator; also on how clean it is. Surface leakages will reduce the current output, especially in warm, humid weather.

The maximum current occurs at about halfway between maximum and zero voltage. The maximum, for Curve a, is around 8 microamperes. A microampere is a millionth of an ampere. At half-speed, it would be about 4 microamperes.

Separate Excitation

With separate excitation, high voltage can be maintained on the inductors, they no longer being connected to the collectors. Therefore, as when some demonstration device is connected to the terminals and thereby lowers the machine's output voltage, induction charging of the rods is *maintained*, and current at lower voltages will be anywhere from somewhat greater, to much greater. For example, if we have two Dirods, and No. 2, running unloaded and having only to maintain inductor voltage at maximum of No. 1, the characteristic of No. 1 is Curve e, Figure 7.

Separate excitation can be had without using a second Dirod. Instead, an exciting capacitor is mounted above the Dirod, Figure 8. First, the capacitor is connected to the inductors. Second, the inductors are temporarily connected to the collectors. Running the Dirod soon charges the capacitor. Inductors and collectors are next disconnected.

With separate excitation, the generator polarity is not reversed when the generator is shorted.

The Capacitor

My capacitor is of 1/8-inch Plexiglas, with a round plate on each side, epoxied to it. The plates, of 6-inch diameter, are 1/16 inch thick, with edges rounded and smoothed. The edges are heavily coated with silicone rubber, laid out on the Plexiglas for a quarter inch or so, and up over the plate rims by as much. Even thus protected, incipient discharging is seen at above 60 kV, and thus its voltage limit is about that value. If thus charged, the charge gradually leaks away. The characteristic curves for capacitor voltages of 20, 40 and 60 kV are, in order, Curves b, c, and d. *This capacitor is safe.* I have accidentally had its full discharge to my hands many times. Not really painful, but unpleasant.

Instead of silicone rubber, Corona Dope can be liberally applied. If 1/16-inch aluminum stock for the plates is unobtainable, cut heavy aluminum foil,

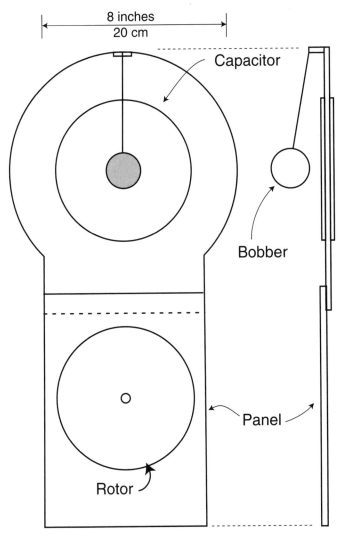

Figure 8. Dirod with capacitor

roll it flat, and use it. The capacitor Plexiglas extends an inch below the panel, and is attached there by two screws.

5. Making the Accessories

The Bobber

A ping pong ball, painted with TubeKoat, hangs by a thread from the projection shown, so that it just touches the front capacitor plate. When I first installed it, I had only one thing in mind: I hoped that it would, on being charged, be repelled by the plate, stand out somewhat from it, and visually tell me the capacitor was charged. It did. And then, to my surprise (and pleasure!) it started bobbing, swinging in and out. Any audience finds this fascinating to watch. I explain it thus: on touching or nearly so, it is fully charged; in the time it takes for it to swing out and in, enough charge has climbed the slightly conductive cotton thread and dissipated by unseen corona on the thread, to reduce the charge somewhat, and the repulsive force. Every so often, electrostatics likes to hand us a pleasant surprise. If the thread is too dry, it won't bob. Moisten the thread.

Plugs and Receptacles

The two bent rods earlier taped on to connect inductors and collectors are now removed and discarded. Receptacles, Figure 2 and Figure 9, each of two pieces of Plexiglas, are epoxied onto the inductors, making slots 1/8 inch wide. The plugs, Figure 10, are now contrived. Mine are of 1/16-inch aluminum, half an inch wide, edges rounded and smoothed. The one at the left has its upper half bent a bit, to make it hold firmly in the slot, and not slide out. Yours can be of 1/8-inch rod, attached to insulating handles of your choice.

Connectors: Capacitor Plates to Inductors

Flexible rubber-covered wire is needed for this job. TV repair shops should carry two kinds of Belden wire, called Test Prod wire: one guaranteed for 3000 volts, the other for 300. Either kind will do, but the second kind is preferred, being more flexible. (Those voltages have to do with human safety). Actually, these wires give fine service for far higher voltages. Cut two pieces of wire, amply long. End each with a plug (see Figure 11). Plugs can be 1/8-inch rod; or if Bake Rod, filed down to plug into the inductor's receptacles. Plugs to be rounded at ends, and smoothed all over. Use thread wraps to tie wire end to plug base. Wrap tightly with a strip of glass tape. Paint the wrap lightly with a first coat of epoxy. Let set. Then give a heavier coat, to make

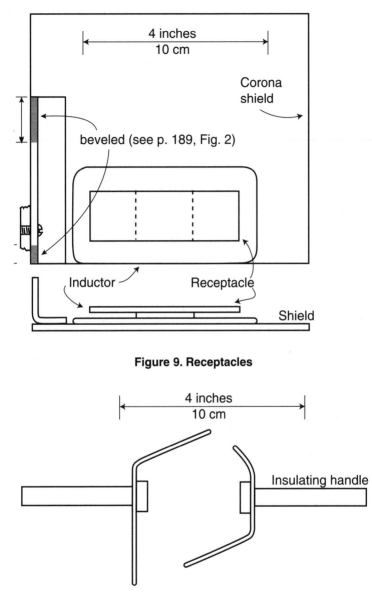

Figure 9. Receptacles

Figure 10. Plugs

the joint rigid. Now plug the plugs into the receptacles. Carry the left wire up and around, then lay it against the face of the front capacitor plate so that it lies roughly along a radius. Cut the wire to be about halfway along the radius.

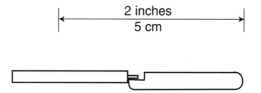

Figure 11. Plug ends

Cut the wire, and strip a half inch of rubber off. Apply a patch of tape to cover wire and insulation. Tape tightly all over; again use a light epoxy coat, plus a heavier coat. Do likewise for the right wire. The reason for a first light coat is this: a heavy coat at first just might manage to creep in and insulate the wire from plug or plate.

Operational Flexibility

We now have a machine with maximum flexibility. With all plugs in, the capacitor can be charged. Leaving all plugs in, there could be some experiments that work best that way. Remove the plugs that tie collectors to inductors, and have a separately-excited generator. Or, leave those in, unhook the capacitor, and we come back to the self-excited generator. You will be experimenting with all of these choices, and finding what is best for a particular demonstration.

More about Capacitance

The unit of capacitance, the farad, is enormously large. Smaller units are the millifarad (1/1000 of a farad); the microfarad (a millionth of a farad); and the picofarad (a millionth of a millionth of a farad) and pronounced "peeko" farad. Its symbol is pF. A Dirod has some capacitance: something like 5 pF. My exciter capacitor, having 1/8-inch Plexiglas, has about 160 pF.

A lot of capacitance can be very easily achieved, simply by using several feet of two insulated flexible wires and twisting them together. Take two 7-foot lengths of Test Prod wire, twist 5 feet (leaving separation at both ends). Since the wires are so close together, the electric field between them is very intense, and charge density on the wires is very high. This capacitor may be as high as 60 pF. Once in Chicago, demonstrating at a convention banquet, I wanted to set off good sparks at a spark gap on a center table, but with the Dirod up on the platform. I used maybe 50 feet of each wire, not even twisted, but merely tied together every so often. With lights out, the intermittent sparks achieved were highly satisfactory.

This may lead you to think that instead of troubling to build your Dirod's exciter capacitor, you could easily get the same capacitance by twisting a couple of wires. You could. But due to that same very intense field between the wires, the charge would leak away much too fast.

Two Kinds of Connectors

Rigid

One is shown on Figure 6. Two are needed, about 10 inches long, to connect the Dirod to the demonstration item. Mine are 1/4-inch aluminum. To suppress loss of current by corona, I have turned brass endings for them, as indicated by the shaded outline. You may get by very well without the endings. But if you have to use 1/8-inch brass rod for connectors, they certainly will require enlarged metal endings; or, capping with Plexiglas endings fitted down over, and epoxied on. Another way out is to use two pieces of Bake Rod for a connector—placed end to end, and pieced together by enough wraps of glass tape, coated with epoxy for added stiffness at the joint. They will need end-capping, too. Do not overlook 3/16-inch copper tubing, easily bent to your needs. Such connectors will certainly need end-capping to reduce corona.

Flexible

I make a lot of use of my flexible leads, made of two lengths of Test Prod wire, each about 20 inches long, and terminated with pull-chain loops. See Figure 12. Cut the exposed wire strands just long enough to wrap them tightly around the link between the two balls. Wrap the joint all over with a strip of glass tape so that when a first *light* coat of epoxy is put on, it will not creep in and perhaps insulate the wire from the chain. When set, add a heavier coat. Finally, at the top where the chain is cut, bring the balls together, wrap with tape, and paint with epoxy.

The Rod Gap: Measuring High Voltage

The simple way to measure high voltage is with a sphere gap. Mine is shown in Figure 13: two stainless steel hollow spheres, 1-1/2 inches in diameter. Such spheres are expensive. (See a copy of the book, *Electrostatics**, for

---

*Part I of this book.

Figure 12. Flexible connector

Figure 13. Sphere gap

further information on sphere gaps.) I have also rigged a rod gap, Figure 14, drawn to show one rod's side view; the other, like it, shown end-on. Diameter, 3/4 inch. My gap had the two axially *in line*, with the spark gap adjustable. The smaller ending on each went through support holes. The sparking ends were finished in the lathe to have quite accurate hemispherical surfaces. Without the lathe, you cannot achieve good surfaces to trust the results. Instead, you can convert the pair into the *crossed-rod gap*. At their middles, the two are at

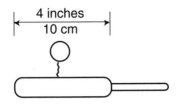

Figure 14. Crossed-rod gap

two are at right angles, with the gap between made somehow adjustable. Now, my rods (the large parts) are only 4 inches long. Too short for rod-gap use. The spark region would be too much affected by the nearness of the electric field effects of the ends. Your length should be 6 inches. Also, using flexible leads, attach these to the ends of the small inserts, and do not let a lead be close to the spark region. The calibration is shown in Figure 14. If you cannot

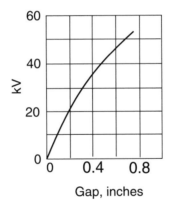

Figure 15. Spark gap calibration

get 3/4 inch solid bar stock, use brass pipe of that diameter, and very effectively cap the ends with Plexiglas, epoxied on. Copper tubing could also be used, properly capped against corona.

6. The Demonstrations

Preliminary Demonstrations Notes

In presenting 25 demonstrations all gathered together, the impression may be given that I developed them all rapidly, and in some orderly manner. The truth is utterly different. My large laboratory is also my office, and I am in it seven days a week. Thus, when an idea comes, I can immediately put it to trial. If it works, fine. If not, it is either improved and adopted, or given up. Thus, these demonstrations were worked up over the years. And importantly, some "invented themselves." Electrostatics likes to play tricks. In trying an idea experimentally, it sometimes produces a surprise, and, a new effect is born. *Do not hesitate to try things.* You may be surprised!

Some of the demonstrations are shown complete. Others, not complete, show only the electrical necessities. For these, you are on your own as to completing their structures. There are two reasons for omitting these structures. First, I lack the time to do the drawings. Second, my items, all of which I designed and made myself, must be designed for the lecture circuit. They must pack in minimum space; be ready for making a fast setup for a lecture; and often safely packed again for transport in minimum time. Yours are not subject to such requirements.

Materials

Kneeling pads

These are carried in the dollar stores, for housewives and worshippers (I suppose) to kneel on. Mine are 8 by 9 by 2.5 inches. When you demonstrate some items you may need a table surface of high insulative quality. Don't count on it; and, some tables have metal tops. In a Science Fair you may have to use the table provided. Two kneeling pads should be part of your kit.

Styrofoam slabs

Styrofoam, firm and stiff and pure white, is now widely available. Lots of it is used by firms for packing and protecting shipments of instruments and other things. Much of this stuff is thrown away. Dollar stores should carry it in some form.

When you need to isolate yourself from ground, you will stand on such a slab. A slab a foot square and about an inch or more thick will be adequate.

Little balls

The supermarket carries little balls, used for trimming cookies and cakes. *Decors* are hard sugar balls, silver-coated, roughly 1/16 inch in size. They are used in Figure 29. Another kind, *Dragees*, are larger. *Trimits*, (also called dragees) are still larger. Try experimenting with any of these. *Nonpariels*, smaller than Decors, are red and green sugar balls, uncoated. A mix of these with Decors is used in Figure 30.

Ping pong balls.

These are to be conductively coated with TV Tubekoat.

Needles for shafts.

Some electrostatic motors can best be made with needles for shafts. To cut off a needle to desired length and discard the pointed part presents a problem. If it is a good, high-tempered needle, you cannot file it through. The edge of a grinding wheel can be used. Or, break it off. Clamp it in a vise, *point down*. Place something hard against it, touching at the surface of the vise jaws, hit it a sharp rap with a hammer, and snap it off. To clamp it, point up, is to risk an eye. The piece might by rare chance bounce on something, and hit you. *Always wear safety glasses when doing this kind of work.*

Dirod Behavior when Loaded

Dirod characteristic curves, Figure 7, have been briefly considered. Let us go further. With nothing connected to the terminals, the Dirod is *unloaded*. Any demonstration item connected to it is properly called a *load*. Many of the items have their operating voltages indicated, such as (20 kV), (60 kV) and so on. This could lead you to think that when you are about to connect a given load, you would first adjust the Dirod's voltage to the load's preferred voltage. Actually, you will not, and cannot, do that. Immediately after disconnecting one load, the Dirod climbs quickly to its maximum voltage, around 80 kV. Then you connect the next load (the next demonstration item). The Dirod drops its voltage, often to something near to what the load "likes," to operate at its best; but, depending on rotor speed, the load may be getting too much or too little current. Whereupon, you adjust the speed (and the current output at the same time) for best behavior for that load. For example, levitation (Figure 35) may require close Dirod speed adjustment to get the floater to behave.

There are exceptions. For example, in running a motor, such as Figure 22, we have a load such that at very low generator speeds, the motor speed is very low; on up through a great range, the higher the Dirod speed, the faster the disk runs.

Polarity Indicator

The clear plastic bottle (Figure 16) came full of Heinz vinegar, bought in

Figure 16. Polarity Indicator

Canada. If the U. S. version continues to be glass, look around for some other clear plastic bottle. (No clear plastic drinking glass I have tried gives good results.) Put in enough Decors to cover roughly 2 square inches, in a monolayer. Shake the balls around. What used to be called frictional electricity, and now called triboelectric effect, occurs. The bottle becomes negative. The balls, positive, are attracted to the bottle, but repelled by each other. They crawl around.

Hold the bottle to a Dirod collector. If the balls are attracted, the collector is negative. If repelled, it is positive. My first two Heinz bottles only needed cleaning with soap and water, rinsing, and drying, to perform. The third did not work well until cleaned with hot Drano solution. Surfaces become contaminated with all sorts of things, and some greatly modify the triboelectric effect.

Cups that Repel

Two Styrofoam coffee cups, hung on a thread (Figure 17). Charge a cup inside and out, rubbing with a clean cotton handkerchief. You may find some other cloth as good or even better. Set it down. Charge the other cup. Suspended, mutual repulsion of like charges is demonstrated. They induce the opposite kind of charge on you, and are attracted to you. You are demonstrating why dust sticks to walls. (Virtually all dusts are charged.)

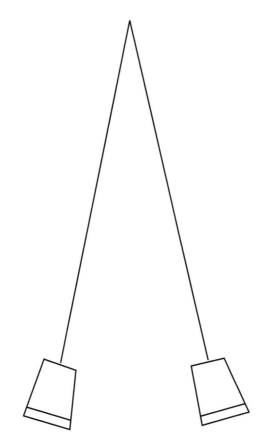

Figure 17. Cups that repel

The Ping Pong Pair

The coated balls are suspended (Figure 18) by fine brass chain (from the dollar store). Chain attached at top to a short rod so that with, balls uncharged, they hang, just in contact. Ground one Dirod terminal. (Grounding to a metal table, or steel underworks of a table, will suffice. If you use a Variac or Powerstat for rotor speed control, ground to it.) Stand on a Styrofoam slab to isolate yourself. Hold the ball's rod by one hand, connect the other hand to the live side of the Dirod. You will be charged up to maybe 50 kV. The balls swing several inches apart, by mutual repulsion. Next, release your contact with the Dirod, and watch the balls. Their slow approach to each other will show how well you and the chain and balls are losing charge.

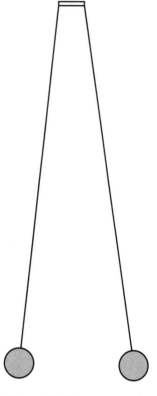

Figure 18. Ping pong pair

Water Spray

Mutual repulsion again shown. Get an ear syringe at the drugstore. Cut off the tip at the right place to let a tube be inserted with a fine orifice. When it all works right, you can fill the bulb and send a long spray for 10 feet across the room (Figure 19). Uncharged, the spray hangs pretty much together, even though it is actually breaking into drops soon after leaving the orifice.

Again isolate yourself on the slab as above, ground a terminal, get charged up from the other terminal, and squeeze. The charged drops now repel each other, to make a wide spray.

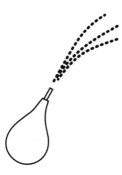

Figure 19. Water spray

Clatterbox

"We stand on the shoulders of those who have gone before." This time, it is Benjamin Franklin's shoulders. After inventing the lightning rod and putting one on his house, he brought the conductor inside (which he should not have done!) and connected it to a bell. Nearby, another bell which he grounded. In between he hung a metal ball, on a thread. When a charged thundercloud came overhead, the ball clattered back and forth and rang the bells. My "bells" are two flat tin cans, with lids on (Figure 20). They make a fine racket.

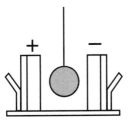

Figure 20. The clatterbox

Yours can be two 3-inch discs, 1/16 inch thick, corners rounded, edges smoothed, to reduce corona. If unavailable, use Plexiglas discs, Tubekoated all over. The charged ball is driven across by the electric field between the discs; touches; has its charge reversed; is driven back, and so on.

Two-ball Clatterbox

A very interesting extension of the above. Structure, all Plexiglas, except for the flat electrode uprights (Figure 21). These are as above, aluminum or

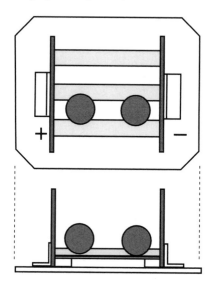

Figure 21. The two-ball clatterbox

coated Plexiglas. *Do not* reduce the size (4 by 6 inches); the balls may become erratic. In Figure 20, Plexiglas strips attached to the cans accept the connectors from the Dirod. The same can be used here (not shown). The three floor strips, 1/4 inch thick, are spaced to make two ball tracks, so that the balls roll on the bottom, with strip edges nearly touching them. With *one ball*, operation is the same as above. With *two balls* in one groove, they bang together in the middle, discharge to each other, separate to bang the electrodes, get charged (oppositely), collide again, and so on. With *one ball in each groove*, you get alternate synchrony. The balls pass in the middle, one going east, the other west. To understand the two-ball cases, don't forget that the balls attract each other. Everyone likes this Clatterbox. (Practice the operation by adjusting Dirod speed to get the best performance.)

> (Coming next, a series of motors, all operated by corona discharges from blunt electrodes putting charges on insulants that rotate. Electrodes are 1/4-inch aluminum rods, both ends rounded and smoothed. If not available, use Bake Rod material.)

Lid Motor

Rotor, the 4-inch plastic lid used to cover a one-pound can of coffee after it is opened (Figure 22). Pierce an exact center hole with a needle. Rig a needle

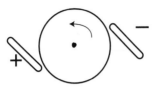

Figure 22. The lid motor (45 kV)

for the shaft. It is self-starting, with those bias electrodes. The electric wind from the electrodes helps to turn the disk. But the major torque is due to charges on the disk rim being urged by the electric field of the two rods.

Tri-Motor

Plastic lids again, 3-inch (or adapted to any other size). Mine are from Pringle's New-fangled Potato Chips. If accurately centered and given good bearings, very high speeds are attained. The rectangles shown are thin metal plates (Figure 23). Underneath, in mine, blocks of Plexiglas, drilled to receive

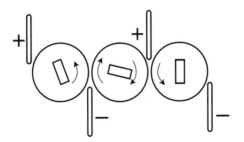

Figure 23. The tri-motor (60 kV)

blunt-end type needles. The top plates rest on the blunt ends. Epoxy will not stick to polyethylene lids. Therefore, each lid has a hole through it on each side of the bearing. Epoxy is thus able to glue the top plate, through to the Plexiglas piece.

Marble Motor

Three glass marbles, 7/8-inch diameter, are within a brass ring half an inch high (Figure 24). The Dirod connects to the ring and the bent rod elec-

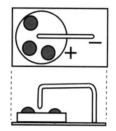

Figure 24. The marble motor (30 kV)

trode. Corona discharge sprays the marbles, and for the most part, onto their nearby surfaces. If the marbles are given a start, they race around inside the ring, either one way, or the other. If the electrode is plus, the charged areas on the marbles are plus, and these are attracted to the negative ring.

Cup Motor

A pasteboard top is taped to the top of a Styrofoam coffee cup (Figure 25). The vertical shaft is an aluminum knitting needle from the dollar store. It rests in a bearing hole in Plexiglas. Your finger and thumb loosely hold the top, to be the upper bearing. The cup rapidly spins. (I also have two other rotors for this: one, a vitamin pill bottle; the other, a clear plastic drinking glass.)

Electric Pinwheel

This goes far back in the early history of electrostatics, and was a matter of high interest for decades. This rotor is a flat strip of aluminum (Figure 26). Another form is to bend a wire at its center to make a wrap around the needle shaft, and bend the ends to point at right angles. If charged by way of a connection to the needle, from one Dirod terminal, the pinwheel spins backward. Since each point produces an electric wind by accelerations away from it, there is a reaction force that does the backward turning. The pinwheel can also be operated with no connection to the generator. Set up two pie pans on edge, facing each other, and maybe 9 inches apart. Charged plus and minus,

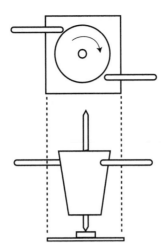

Figure 25. The cup motor (45 kV)

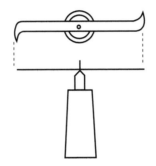

Figure 26. The electric pinwheel

there is an electric field between them. Insert the pinwheel. By induction, one end is charged plus, the other minus. Discharge begins at the points, and it begins to turn. When halfway around, the sign of the charges has reversed, end for end, and it continues. For mine, I have a strip of aluminum foil, from needle shaft to the bottom of the plastic base. Rubberbands attached to the base, and running under my chin, hold the pinwheel on top of my head (where there is now next to no insulation!) I mount my slab, charge myself, the pinwheel turns, and nobody can view that without laughing.

Electric Blower

This is a Plexiglas box, open at both ends (Figure 27). At the left, a point-

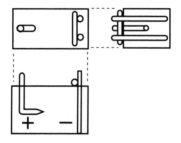

Figure 27. The electric blower (25 kV)

ed electrode. At the right, a couple of rods. These constitute a passive electrode (not discharging). With voltage applied, the pointed electrode makes air stream through the box. Smoke liberated near that end is carried through. Here is a blower with no moving parts. However, do not dream of putting it on the market. The efficiency would be only one percent, or less.

Precipitation

A clear plastic drinking glass with a plastic lid, and a tube inserted through the lid (Figure 28). The lower end of the tube is sharpened, to make

Figure 28. The precipitator

corona. Smoke, put in through the tube, hangs around. Connect the tube to one side of the Dirod. The smoke is churned by the electric wind, then very quickly disappears. Air ions, moving out in the electric field, attach to smoke particles, charge them, and they are moved to the walls and deposited there.

The Hailstorm

The vertical part is Plexiglas, bent into a circle (Figure 29). It is capped

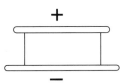

Figure 29. The hailstorm (55 kV)

by aluminum discs, top and bottom. A batch of Decors is poured in. Connect the Dirod to top and bottom. The balls jump up and down, floor to ceiling and back, making a fine active display, and with a hissing sound.

Separating a Mixture

If a mixture of two kinds of small balls or pellets, one conductive, the other nonconductive, is placed in the box that has Plexiglas walls, only the conductives will dance up and down (Figure 30). These will start popping through

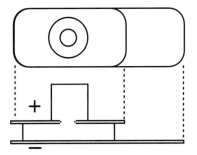

Figure 30. The separator (30 kV)

the hole in the lid, and will be trapped in the glass or Plexiglas chamber above. In health food stores, various seeds are available. You might work up a mixture of Decors (conductive) and seeds (sufficiently nonconductive, if dry enough) to effect a realistic separation.

Popcorn

Foamed polystyrene is now made in great quantities in pieces, very lightweight, soft, a bit springy, and used for packing any number of things. Various shapes (peanuts, discs, etc.) are made. The common name is "popcorn". Pinch off pieces, highly irregular, of 1/4-inch size or more. Cover the bottom of a plastic bowl (Figure 31). Have the bowl on a metal plate (a pie pan will

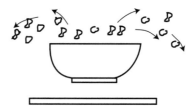

Figure 31. Electric popcorn

do). Using a flexible, pull-chain-ended lead from a Dirod terminal, wave it around, an inch or so about the grains. They soon become charged and show it by shifting a bit here and there. Two more things happen. The bowl's floor surface is also charged. The floor charge would now repel the popcorn and throw it out, were it not for another charge: the lower plate has become charged by induction, opposite in polarity to the upper charges. Suddenly lift the bowl. The "popped" popcorn flies up and out in all directions. Very effective!

Franklin Motor

Franklin made a 28-inch motor, using glass arms, and with thimbles on the ends for charge-carriers (Figure 32). Your carriers will be coated ping pong balls. The central part is a Plexiglas disk, mounted on a bearing on top of a post. Two electrodes, plus and minus, are mounted about as far apart as two adjacent rotor balls, and close to the balls as they rotate. This is a spark device. The electrodes can be two coated ping pong balls. They charge the passing balls, not by corona, but by sparking to them. A passing ball between the electrodes is moved along by the electric field between electrodes.

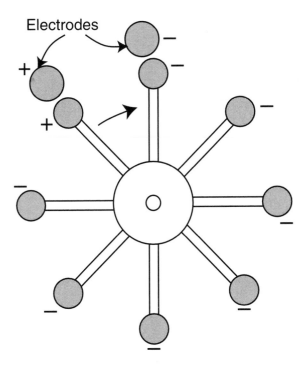

Electrodes

Figure 32. The Franklin motor (60 kV)

Field Indicator

The base is a Styrofoam slab, 3/4-inch thick or more. Twenty pins are stuck into it, arranged in arcs of circles (Figure 33). Each pin carries a piece of 3 by 5 card, pierced, to turn easily on its pin as a shaft. The pieces should be colored dark or black, to be easily seen. In the front view a little circle is shown at the middles of the sides. These are end views of 1/4-inch rods, 16 or 18 inches long. The electric field between parallel rods does have the field shape the pieces will portray. If the pins are helter-skelter at first, apply high voltage to the rods. The field acts to line up the pieces with the field lines, when they flip into position. The effect is striking, even for experienced workers in science or engineering. (To hold the pieces out near the pin heads, slide a small bead onto the pins and hold with epoxy.)

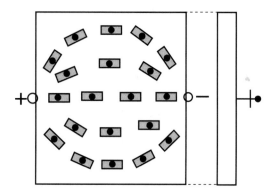

Figure 33. The field indicator

Leyden Jars

Made of a vitamin pill bottle, having a screw cap (Figure 34). The outer

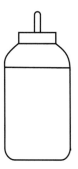

Figure 34. Leyden jars (35 kV)

aluminum foil, applied to overlap itself and taped or epoxied on, is easily applied. Make it go no higher than is shown. The inner foil's application gave me a very bad time until, in desperation, I thought of using electrostatics to get it nicely placed. Cut the inner foil to match the outer one, and to overlap itself somewhat. Roll it flat. Then bend it around a cylinder to curl it enough to slip in through the bottle mouth. Manipulate it into place, with the bottle on its side. Now have the middle of the foil down, at the bottom. Connect one Dirod terminal to the outer foil. Have some kind of rod with an insulating handle; connect it to the other terminal. When it touches the inner foil at bottom, the foil immediately opens out and smoothly flips into place, ready for taping.

To connect to it, puncture the cap, and carry a piece of rod down through it. Let a piece of pull-chain dangle from it, into crumpled aluminum foil pieces placed inside to make contact with the foil. When I gang up my three Leyden jars of this size, and discharge at around 40 kV, there is a fine loud spark. Though not dangerous, you would never take it a second time. Too painful. I also have a gang of three smaller bottles. These jars are very tough, very light, very cheap.

Levitation

Benjamin Wilson, an English associate of Franklin's, somehow discovered levitation, probably in the 1750's. Below, a pizza pan (Figure 35). Sus-

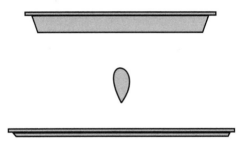

Figure 35. Levitation (30 kV)

pended above, an 8-inch pie pan. Between, a vertical electric field. Cut a piece of foil from aluminum foil; or from metallized paper, such as that used for a package of cigarettes. No matter how you vary the shape otherwise, to make a little man, a fish, etc., all floaters must have a *rounded top*, and a *sharp foot* (or feet). *Have the top pan negative.* (No one knows why it won't work the other way). Reach in, release the floater, and if all goes well, it levitates! The mix of phenomena present make this a very complex thing to analyze; so much so that it may never be completely analyzed. Sometimes you need to play around with Dirod speed, and with the vertical pan spacing, to get results. Floaters are jealous. Only one at a time, will work.

The Jumper

Cotton thread, when clean and very dry, is a good insulator. Wetted, it becomes quite conductive. In between is a great range of semiconductivity. Make a tangle of thread, a somewhat self-supporting tangle reaching in all directions for maybe two inches. Lay it on the lower pan (Figure 35) and turn on the voltage. If very dry, the thread does little: charge is very slow in creeping into it. Remove it, use the moist breath to raise conductivity. Gradually, the strands will begin to stand up. Breathe on it some more. It will unfold more, rise higher. More yet, and it starts to jump from pan to pan. Fun to watch.

The Rockers

My big rockers are of 1/16-inch aluminum, an inch wide, and weighted to stand as shown when at rest (Figure 36). My capacitor, of 1/8-inch Plexiglas, has a large aluminum plate on the under side, and two smaller ones on top. Connect Dirod to the upper plates. As the capacitor charges up, the rockers swing together, and nicely spark to each other before banging together. Now they rock apart. As they do so, more capacitor charging occurs, they swing together again, and so on. Everyone likes this show. (For giving my show in England, and going by air, I made little rockers, about half-size all over.)

The Ball Race

This action-filled attention-getter has coated ping pong balls racing around inside a Plexiglas fence one inch high (Figure 37). The fence, heat bent to shape, rests on Plexiglas a foot square. At each side, two elevated balls serve as electrodes. The spark gaps between these and the racing balls is 3/16 inch. The balls are fed in, one by one. It is easy to race six balls, and seven. At eight, crowding may begin, but nine may be successful. Ten, often possible. Twelve, very rarely possible, is the limit.

This demonstration is a fine example of the persistence it sometimes takes to develop an electrostatic phenomenon to where it is reliable. Over a long period, I would at times get reliability; other times, misbehavior. At last, I discovered that it requires the right kind of base. Instead of a dense polystyrene slab sometimes used, it requires the much lower density furnished by the kneeling pad.

"Perpetual Motion"

Here is a great mystifier. The pizza pan of Figure 35 is turned upside down. On it rests a sheet of Mylar, ten mils (0.010 inches) thick (Figure 38). You start a coated ball around the track. It gathers speed, races, and keeps on going, and going, and going, typically making 80 to 120 turns or so before

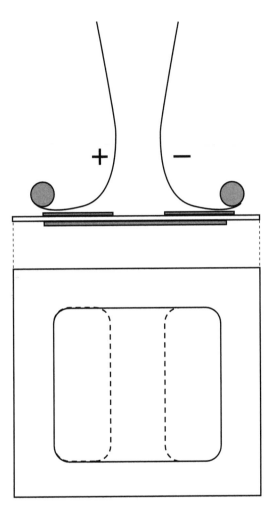

Figure 36. The rockers (40 kV)

stopping. Once, mine made 631 times around! Often, I start my show with it. Before it begins, I do the charging, using the flexible leads with pull-chain ends, dangling the two ends and letting them slide all over the areas, plus and minus, bounded roughly by the dotted outlines. These are *very strong charges*. They induce equal and opposite charges on the top of the pan. After charging, if you were to put your hands down in, you would get a bad shock. Not harmful, but painful. The ball runs over the plus area, gets charged plus. There is an electric field between the two areas, and that impels it to keep going around to the minus area, and so on. The action is so strong that I can carefully

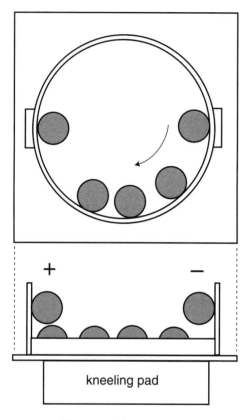

Figure 37. The ball race

pick up the assembly at the neutral region, tilt the whole thing, and have the ball racing uphill. To completely explain everything about it all, requires five sketches and two typed pages.

Tassels Terrific

Here is the loveliest demonstration of them all. Two academic tassels (used on the mortarboards worn at graduation) are hung from the outer ends of the two horizontal rods used in Figure 33. I use gold and blue, the University of Michigan colors. My rods are mounted to swing out or in. Parallel, they are somewhat over a foot apart (Figure 39). The tassels are only slightly conductive. (In cold dry winter weather, they may need to be hung in a chamber with a bit of water at the bottom, to give best performance.)

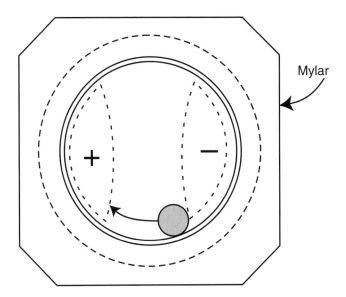

Figure 38. Perpetual motion

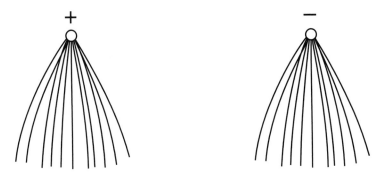

Figure 39. Tassels terrific

A means of easily attaching the tassels to the rod ends must be devised. Hunt around for a tube to slide onto a rod (see details, Figure 40). It can be metal, or insulant. It slides on until the rod touches the metal band around the top of the strands. Tape the tube to this band with glass tape. Next, a corona guard must be fitted over each such (shown at the right). It can be a band of 1/16-inch aluminum, edges and corners rounded and smoothed; or, of heat-bent 1/8-inch Plexiglas, Tubekoated all over, inside and out. Without guards,

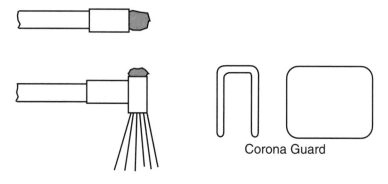

Corona Guard

Figure 40. Details of tassel mounting

the sharp edges of the bands could spill most of the Dirod output, leaving little to flare out the tassels.

With the Dirod connected, and hanging apart, each flares out. Move a hand close to a tassel, and the strands swing to approach and cling.

Swing the two together somewhat, and their strands beautifully interlace with each other. It is interesting, then, to turn off the Dirod. As the charges are slowly lost the strands intermittently drop apart, until only one or two pairs are still clinging, joining tassel to tassel.

Next, with tassels maybe 16 inches apart, light a candle. *A candle flame makes ions of both kinds*, plus and minus, *in equal numbers*. There is an electric field between the two tassels. If the flame is held halfway between, the negative ions should go to the positive tassel, and the positive ions to the negative tassel. They do. The tassels equally wilt.

Finally comes one of the finest demonstrations of ions you will ever see. Swing the tassels wide apart. Take a point (an ice pick works, and can be seen by all) in one hand. Isolate yourself, standing on the Styrofoam slab. Stand to one side, to be removed from the far tassel. Charge yourself by grasping the near rod. Point the ice pick toward the audience. Ions are streaming from it, but largely directed along the strong field of the point and toward the audience. And now, with your hand five feet from the far tassel, turn the pick to point toward it. Ions, now streaming toward it, will cause a little wilting, even that far away.

Spray Painting

My target is a 3 by 7-inch aluminum plate, 1/16-inch, corners rounded, edges smoothed. A long lead goes from one Dirod terminal to it. From it goes a string held high by an assistant, to dangle the plate in space, and make sure no paint will get on nearby apparatus or such. I stand on the Styrofoam floor slab, with a spray can of Krylon. With the can somewhere around two feet from the target, I first aim the cloud of spray about four inches from one side of the target. I am not as yet charged. Only a few dots of paint will hit the target. Next, I hold the other terminal, get charged, the can and the paint is charged, and a bright band of color about an inch wide gets onto the target. Moreover, the rear side typically gets a wider band!

7. A Final Word

Electrostatics is endlessly fascinating, and I repeat what I said on page 174—"There is more fun in electrostatics than in any other area of science." Once you get your Dirod to operating, the door is wide open for your own experimenting. You may discover demonstrations in addition to those described, and thus make your own contributions.

Earlier, I mentioned some of the major *applications* of electrostatics. Since the Manual is concerned with how to build and operate your equipment, it has made no mention of another side to electrostatics: its hazards.

Electrostatic Hazards

These are many, and some are serious indeed. Lightning, of course, comes to mind at once. It is spectacular and can be deadly.

Airplanes, flying through dust or snow, can become highly charged. In particular, helicopters, hovering, can stir up dust, and the blades whirling through dust can build up very high charges on the helicopter. If a hook is lowered to pick up something from a deck, a man reaching for the hook can suffer very severe shock.

Colossal explosions occur when the dust in grain elevators explodes, wrecking those great silos and killing men. Sometimes the ignition of the dust comes from sparks made by the charged dust itself.

Oil supertankers have blown up. In December of 1969, three new supertankers of 200,000 tons or more, blew up in three different parts of the ocean. In each case, they were being "washed down." Gunk tends to accumulate on the walls of a hold. To cut it down, giant water sprays were used. When water jets struck the walls, billions of water drops were formed, and these are charged. So, here was, in effect, a highly charged cloud. Almost certainly, it was a spark caused by such a cloud that ignited the vapor in the hold, blowing up the ships.

Industry is busy all the time, engaging in new processes, or perhaps speeding up old ones. Triboelectric (frictional) effects may develop, at least causing nuisances, or at worst, making discharges causing fires or explosions.

The literature on electrostatics is enormous, and is growing all the time. Much of it comes by way of papers given at conferences held each year here, in the United Kingdom, in Europe, and in Japan.

Materials List

(After writing the Dirod manual, A. D. Moore sent a letter to all those who had requested the manual, indicating his plans to put together a kit to aid in construction of the Dirods. Following is an excerpt from that letter along with the list of materials that he suggested; it is included here in the hopes that it proves useful to experimenters.)

Ann Arbor, Michigan
February 5, 1981

To those who have ordered copies of the Dirod Manual:

Herewith is your copy, and attached hereto is your brush material...

The finding of materials needed for the Dirod and Demonstration items may be considerably eased for you. My materials suggestions are listed below for the Dirod and for the demonstrations.

The dimensions are in excess of finished dimensions. They will come to you rough-cut.

Beyond the Mylar, demonstration materials cannot be forecast here. The needs will depend on which demonstrations will be made up, and on their structural design. The materials and sizes will almost entirely fall within the Dirod list.

To you, good luck and happy sparking.

Prof. Emeritus
A. D. Moore

Table 1. For the Dirod

Material	Thickness (inches)	Number	Length/Width (inches)
Steel shafting	1/4	One piece	15 inches long
Plexiglas	1/8	Two pieces	12 by 12 (or one, 12 by 24)
	1/4	One "	12 by 12 (if the thicker panel is used)
	1/8	"	12 by 12 (for Separate Excitation Capacitor)
Soft aluminum sheet	1/16		7 by 14 (for above capacitor plates)
	1/8	"	4-1/2 by 5 (for inductors)
	1/4	"	4 by 6-1/2 (for collectors)
Half-hard brass rod	1/8	Enough for 36 rods, 4 inches long, & some extra.	
Soft aluminum rod	1/4	One	24 long, or two 12 inch. (Connectors)

Table 2. For the Demonstrations

Material	Thickness (inches)	Number	Length/Width (inches)
Mylar sheet	0.01	One piece for "Perpetual Motion"	14 by 14

Afterword: A. D. Moore
Remembered by his Children

Professor Arthur D. Moore. Arthur Dearth Moore. A. D. Moore. Or, simply, 'A.D.' to his many friends. To us, Daddy or Dad or Pop. His academic, scientific and other achievements have been reviewed in writings elsewhere. We are introducing you to the person we knew. Who, in our view, was an exceptional, outstanding, wonderful, and loving father.

His Youth

The youngest of seven children on a farm in Pennsylvania. By the light of a fireplace, too young to attend school, he read the dictionary. End to end. Ultimately, he developed an extraordinary vocabulary. When we were younger, how easy and lazy it was to have our own walking/talking dictionary around the house.

His ever exploring, inquisitive, and expanding mind resulted in his being skipped grades in a one room school house. He became the youngest freshman to enter Carnegie Tech; and may still hold that record. Several years after graduating he joined the College of Engineering, University of Michigan.

The Young Man

Once settled in, he met a lovely, raven-haired coed who became his wife. When he dated our Mom-to-be she decided to call him 'A.D.' and it stuck to him ever after. It reflected that part of his character which reduced things to the short and simple. The easily said. Easily understood. Direct to the point.

Their honeymoon was unusual.

Dad had an Old Town canoe. He, it, and bride went to northern Michigan and thence paddled down the Au Sable River for a goodly distance; camping as they went; for two weeks. Their adventure started in a drug store in Grayling, Michigan. It had the most delicious double chocolate sodas. For many years, on our way to camping further north, ritual demanded a stop in Grayling to consume at least one each of that remarkable concoction. For Mom and Dad it was an annual celebration of their honeymoon. For us it was an annual celebration for our taste buds.

Camping

Off the beaten track—lovely scenery—camping became a vital part of our family life. Our camping expeditions often took us to a large lake in northern Ontario, Canada. On an island just for us. With sandy beaches. Where we camped year after year. Where a family cabin was eventually built. Where, today, eighty years later, it is still used by us and, in turn, by our children.

Northern Michigan. Perhaps the most astounding experience in our childhood was occurring. Our car loaded to the gills with camping gear. Stopped. On the shoulder of the road. Out of the car. Watching the night sky. The Aurora! Dancing! The magnitude and intensity of colors dwarfing the power of words. An awesome hush but for the distant lowing of frightened cattle. Stupendous curtains of pastel pinks, greens, and yellows pulsating, marching, ever shifting. Momentous!

He had an old typewriter that went with him. Typed with only three fingers; but as fast as any accomplished secretary. He became a prolific writer. Books. Short stories for us when children. Letters that found their way all over the world.

And poetry? A steady stream of verse flowed from his forays downlake. The opening stanzas from "Timbered Hills" are deeply moving to us:

I'm in the mood
For a certain wood
Whose timbered hills are calling

Whose rocks were grooved
When the ice sheet moved
With an ancient force appalling.

The Canadian wilderness, where he loved to fish, and where the fish were loath to bite, offered the kind of solitude necessary for his creativity. While trolling, Dad polished, if he didn't dream up, some of his short stories about "Johnny Scaggs." Johnny was an adventurous youth whose exploits kept us delightfully entertained as youngsters.

Life with Father

He was an accomplished story teller, both written and spoken; sought after as a speaker and toastmaster. More significant for us, he talked to and with us during dinner. His deep and soothing bass voice, his ready sense of humor, his widespread knowledge entertained, entranced and enlightened. It wasn't until we were adults that we realized that he was teaching in disguise. Character building. Problem solving. Recounting important events and scientific advancements. A wealth of information and advice. His interests were so broad that no subject matter, no discipline, no way of life was excluded.

We sang on picnics. When traveling. Camping. Cleaning up after dinner. Dad washed the dishes, and we helped. It's a choice memory because we sang together while doing it; from early on until we children eventually went our separate ways. And the memory of what emerged, when Mom's lovely voice harmonized with his rich bass voice, brings tears to our eyes.

We wonder how Mom managed, in the 1920s and 1930s, to get a male of the species involved in those kinds of chores. Not just occasionally. But regularly. Knowing him, we suspect it was his own decision.

Not to imply that he always behaved in accord with his own decisions. Take chewing tobacco, for instance. Mom made him quit. She believed it inconsistent with the image of a university professor. Sometimes, out camping, he would sneak a wad out of Mom's sight. On these occasions he would demonstrate to us the accuracy of his expectorated projectile; having, in the past, laid claim to being able to hit a spittoon dead center at ten feet.

A farm boy, and in tune with his metabolism, he was early to bed—early to rise. (He made his own breakfast. Usually three poached eggs on toast.) He couldn't wait to bicycle to his lab every day of the week at 4 AM. Two miles each way, six to seven days a week. In rain. In snow and ice. And there were times, particularly in winter, when the bike went one way and he another. Yes. He got hurt. Occasionally. That didn't stop him. Nor did he complain about it.

This, among other things, assisted his maintaining excellent health. He easily built muscle and was much stronger than average. When younger, his sport was wrestling. He did well enough to become the workout companion of an Olymic medalist.

Could juggle five balls with some tricky maneuvers thrown in. He practiced, even on rainy days, on our lawn extension because it was easier to rescue dropped balls from the street than in among the garden bushes.

Even in his eighties, at the cabin in Canada, he persisted in running a chain saw to cut down trees for firewood, along with a daily routine of bathing in the snow-fed lake.

Outside Interests

He was self-confident. With that, and his widespread interests, he found his way into a large variety of activities. He and Mom started and formed a university 'International Night' which continued for a number of years. It was a large student-faculty production, featuring skits representing the various cultures and races on campus.

He sang in the university glee club and Episcopal church choir. Was elected to the town's governing body for nineteen years. Undercover work for the military during WWI. Classified work for the Naval Ordinance Laboratory during WWII.

But of his many and varied pursuits, the mysteries of electrostatics were the most interesting and intriguing. He enjoyed it. It was fun.

Later

When we had grown and left home, he wrote us a daily letter, often two or more pages, encapsulating his experiences, joys, and discoveries of the day. These were exciting letters, for his interests and knowledge were broad. There were excerpts from correspondence with special friends engaged in the frontiers of science. Each letter pronounced that day to be "grand and glorious" regardless of how severe the rain storm, how fierce the wind-driven snow, how frigid, how hot, or how humid the weather.

When others would have complained about a hard day at the office, or high taxes, or any of life's unpleasant or annoying sides, he saw them as challenges. As adventures. As new points of interest. Problems to be solved. Opportunities for new learning. For him, each day was, truly, a beautiful day and filled to overflowing with delightfully new horizons.

If you met him then, you would have noticed the character lines in his face. You would have seen they were carved by smiles—laughter—humor—kindness—caring—optimism. That was the way he was.

Jeanne M. Goodman
Arthur D. Moore, Jr.
Jo. C. Moore

Bibliography

Attwood, Stephen A. *Electric and Magnetic Fields*. Wiley. New York. 430 pages. Perhaps the most lucid text of its kind ever written; the appendix has an excellent historical outline.

Blanchard, Duncan C. *From Raindrops to Volcanoes*. Science Study Series S50, Doubleday Anchor. Garden City, N.Y. 180 pages. A charming and valuable book.

Boys, Sir Charles Vernon. *Soap Bubbles and the Forces Which Mould Them*. Science Study Series S3, Doubleday Anchor. Garden City, N.Y. Reprint of a famous classic of 1890. In making apparatus, he was unbelievably skillful. The accuracy with which he measured the gravitational constant was not improved on for fifty years.

Cameron, Frank. *Cottrell: Samaritan of Science*. Doubleday. Garden City, N.Y. 414 pages. A fine book about a great pioneer.

Dibner, Bern. *Ten Founding Fathers of the Electrical Science*. Burndy Library Publication. Norwalk, Conn. 46 pages. Fascinating little histories of some of the early "greats" and their discoveries.

Encyclopedia Britannica. For general information on electrostatic generators.

Galambos, Robert. *Nerves and Muscles*. Science Study Series S25, Doubleday Anchor. Garden City, N.Y. 158 pages. An engaging description of the incredible electrical networks of the human body; EKG and EEG waves; electrochemical mechanisms of muscle and nerve cells.

Loeb, Leonard B. *Electrical Coronas*. University of California Press. Berkeley, Calif. 694 pages. This is *the* authoritative work. An advanced treatise.

MacDonald, D. K. C. *Faraday, Maxwell and Kelvin*. Science Study Series S33, Doubleday Anchor. Garden City, N.Y. A delightful book about three splendid men. Be sure to read it!

Miller, E. P., and Spiller, L. L. "Electrostatic Coating Process" (Part I and Part II), *Paint and Varnish Production*, June, July, 1964.

Murr, Larry E. *The Biophysics of Plant Electrotropism*. Transactions of the New York Academy of Sciences, May, 1965. Refers to five other papers by Murr.

Pierce, John R. *Electrons and Waves*. Science Study Series S38, Doubleday Anchor. Garden City, N.Y. 226 pages. An excellent book by one of the master minds in electronics.

Pohl, Herbert A. "Nonuniform Electric Fields," *Scientific American*, Vol. 203, page 107. Much of interest here; striking experiments with liquids.

—"Some Effects of Nonuniform Fields on Dielectrics," *Journal of Applied Physics*, Vol. 29, No. 8, August, 1958. Has interesting experiments; otherwise, theoretical.

Proceedings of the Third International Conference on Atmospheric and Space Electricity. American Elsevier Publishing Co. New York. 616 pages. Includes papers on lightning, lightning rods, ball lightning.

Ralston, O. C. *Electrostatic Separation of Mixed Granular Solids*, American Elsevier Publishing Co. New York. 261 pages. The one book of its kind. Long with the U. S. Bureau of Mines, Ralston was an early Cottrell associate.

Robinson, Myron. "A History of the Electric Wind," *American Journal of Physics*, Vol. 30, 1962, page 366. Excellent historical account; has a long bibliography.

—"Movement of Air in the Electric Wind of the Corona Discharge," *Communications and Electronics* (AIEE), May, 1961. Report on his very extensive research; considerable amount of theory; his Fig. 1 shows an electric blower.

—"The Origins of Electrostatic Precipitation," *Electrical Engineering* (AIEE), September, 1963. An absorbing historical review; has a long bibliography.

Sweet, Richard G. "High Frequency Recording with Electrostatically Deflected Ink Jets," *Review of Scientific Instruments*, Vol. 36, No. 2, February, 1965.

White, Harry J. *Industrial Electrostatic Precipitation*, Addison-Wesley. Reading, Mass. 376 pages. White was with Research-Cottrell for years; the only book of its kind.

Williams, L. Pearce. *Michael Faraday*. Basic Books. New York. A very comprehensive biography.

Index

A Source for all Areas of Electrostatics

Laplacian Press and The Electrostatic Source catalog bookstore supply books and video tapes to support those with interests in all facets of electrostatic phenomena.

Laplacian Press

Laplacian Press publishes new and out-of-print titles, as well as the proceedings of conferences held on electrostatics. It also reprints the classics such as this present volume, A. D. Moore's *Electrostatics, Exploring, Controlling, and Using Static Electricity,* and L. B. Schein's indispensable book for printer and copy specialists, *Electrophotography and Development Physics, rev. 2nd edition.*

The Electrostatic Source

The Electrostatic Source, a catalog bookstore, provides a source for new books, recent technology, and hard-to-find titles in all fields of electrostatics including: ESD; Fundamentals; Coatings, Sprays and EHD; Xerography and Displays; Dielectrics; EMC and Hazards; Bioelectricity; Electrostatic Math; Laboratory; Corona and Gas Discharge; and Popular Works. For a catalog with detailed descriptions of over 100 listings, contact the Electrostatic Source at (408) 779-7774 or email to electro@electrostatic.com.

Electrostatics on the World Wide Web

Connect to the website at http://www.electrostatic.com for online information of the latest book releases, conferences, organizations, applications information, and other websites focusing on electrostatics.